The Income Factory Guide

Generating Cash Flow with Options and Dividends

Joseph Trek

Published by NobleTrex Press

For permissions and other inquiries, write to:

P.O. Box 3132, Framingham, MA 01701, USA

Contents

Introduction

Let's start with a scenario that, on the surface, seems completely absurd.

Imagine a business owner named Sarah. Sarah owns a successful manufacturing plant that produces high-quality industrial tubing. Her factory runs twenty-four hours a day, shipping out thousands of units and bringing in a steady, reliable stream of cash every single week. She pays her employees, covers her overhead, and takes home a handsome profit that pays for her mortgage, her vacations, and her retirement savings.

Now, imagine that every morning at 9:30 AM, Sarah walks out to the parking lot, stares up at the brick façade of her building, and frantically asks a passerby, "How much is this building worth right now?"

If the stranger says, "About two percent less than yesterday," Sarah collapses in despair. She feels poor. She panics about her retirement. If the stranger says, "Three percent more than last week," she feels euphoric and considers upgrading her car.

She does this every single day. She is obsessed with the liquidation value of the bricks and mortar, ignoring the fact that the machines inside are humming along, printing money regardless of the real estate market's mood.

You would call Sarah insane. You would tell her that the value of the building only matters on the day she sells it. Until then, the only thing that matters is the *production*-the income the factory generates.

Yet, this is exactly how 99% of investors treat their portfolios.

We have been trained to function as museum curators rather than factory owners. We collect assets-stocks, ETFs, mutual funds-and we put them behind glass. Then, we spend the rest of our lives staring at the price tag, hoping it gets bigger. We judge our financial health by a single, flashing number at the bottom of a screen: Net Worth.

But here is the uncomfortable truth that the financial industry rarely whispers: You cannot pay your electric bill with Net Worth. You cannot buy groceries with capital appreciation. You can only spend cash.

If you have a million-dollar portfolio but it yields 1%, and you need $50,000 a year to live, you are forced to become a liquidator. You have to sell off pieces of your "factory" to survive. You are praying that when you need to sell, the market is in a good mood. If the market crashes the year you retire, you are cannibalizing your assets at the worst possible moment. That isn't financial freedom; that is financial fragility wrapped in a high balance.

This book is about firing the museum curator and hiring the factory foreman. It is about a fundamental shift in how you view your money. We are going to stop obsessing over the resale value of the building and start focusing on the output of the machines.

Welcome to the Income Factory.

The Two Engines of Production

The philosophy we are about to explore is not a get-rich-quick scheme, nor is it a complex black-box algorithm reserved for Wall Street quant funds. It is a return to the oldest principle of investing: assets should pay you.

However, the "Income Factory" differs from your grandfather's dividend strategy. Traditional income investing usually involves buying utility stocks or bonds and waiting for a 3% or 4% check to arrive in the mail. That is a fine strategy, but it is slow, and in a high-inflation world, it often fails to keep up with the cost of living.

We are going to build a modern factory. Our production floor runs on two distinct engines.

The first engine is **Equity Income**. This is the foundation. We will learn to select high-quality companies that share their profits with us. We aren't just looking for the highest yield-that is often a trap. We are looking for sustainability. We want companies with "moats" so deep that they can keep paying us even when the economy stumbles. This provides the reliable, passive baseload of our cash flow.

The second engine-and this is where things get interesting-is **Volatility Harvesting**. This is where we introduce options.

Now, I can feel some of you tensing up. The word "options" often conjures images of reckless gamblers blowing up their accounts on risky bets, or chaotic screens filled with Greek letters that require a PhD to decipher.

Let me set the record straight immediately: We are not here to gamble. In the casino of the stock market, most people are the players, placing chips on the table and hoping to get lucky. We are not going to be the players. We are going to be the House.

When you buy a lottery ticket, you are buying hope. When you sell a lottery ticket, you are selling probability. The Income Factory is built on *selling*. We sell time. We sell fear. We sell insurance to speculators.

Through strategies like Covered Calls and Cash-Secured Puts, we can manufacture our own dividends. We can generate income from stocks that don't even pay dividends. We can look at a flat market-where the average investor is making zero return-and extract 10% to 15% annualized yields simply by agreeing to buy or sell stocks at prices we effectively choose.

This sounds like magic, but it is actually just a transfer of risk. The market is filled with people who want protection or who want to speculate on a massive rally. They are willing to pay a premium for that privilege. We are the ones who collect that premium. We catch the cash that others leave on the table.

Why This Matters Now

Why write this book now? Because the old rules of personal finance are cracking under pressure.

For decades, the standard advice was the "4% Rule." Accumulate a big pile of money, invest it in a generic index fund, and sell 4% of it every year to live on. That works beautifully in a bull market. But what happens when the market stays flat for a decade? It has happened before. What happens if inflation spikes and bonds-the traditional "safe" asset-lose value alongside stocks? That happened in 2022.

The passive, total-return mindset leaves you entirely at the mercy of market sentiment. If the market drops 20%, you feel 20% poorer. You stop spending. You worry. You check your phone under the dinner table.

The Income Factory creates a psychological firewall. When the market drops, stock prices go down, but *yields* go up. Volatility usually spikes during a crash. Since we are sellers of volatility, a fearful market is actually our most profitable environment.

I want you to imagine a scenario: The S&P 500 drops 10% next month. The news anchors are shouting about a recession. Your neighbor is panic-selling. But you look at your account, and while the *liquidation value* is down, your *income* has actually increased. You just sold a put option for double the premium you usually get because everyone else is scared. You are flush with cash. You are not forced to sell; you are paid to buy.

That is the difference. That is the peace of mind we are chasing. It turns the stock market from a source of anxiety into a utility that services your life.

The Journey Ahead

We are going to build this factory from the ground up, brick by brick.

We will start by bulldozing the shaky foundations of traditional "growth" investing. We need to rewire your brain to stop craving high stock prices and start craving high cash flow. We will discuss the math of solvency and why "average returns" are a lie that ruins retirement plans.

Once the foundation is poured, we will walk onto the factory floor. I will guide you through the toolbox. We will strip away the jargon surrounding stocks and bonds and look at them strictly as income generators. We'll learn how to spot a "yield trap"-a company that pays a high dividend right before it goes bankrupt-and how to identify the "dividend aristocrats" that will pay you for life.

Then, we enter the control room: Options.

If you have never traded an option in your life, do not worry. We are not going to memorize complex formulas. I am not going to lecture you on the Black-Scholes model. I am going to explain options the way you might explain renting out an apartment or insuring a car. We will focus strictly on the strategies that generate income: the Covered Call (renting out stocks you own) and the Cash-Secured Put (getting paid to wait for a discount).

We will demystify the "Greeks"-Delta, Theta, Vega-not as abstract math concepts, but as the levers and dials on your machine. You will learn how to set the machine to "print" and how to adjust the speed when the market gets choppy.

But a factory is more than just machines; it requires management. A significant portion of this book is dedicated to safety. How do you handle a market crash? What happens if a trade goes against you? We will discuss the "Wheel Strategy," a beautiful, cyclical process that allows you to turn losing trades into winning campaigns over time. We will talk about position sizing, because the only thing that can destroy a factory is betting the whole company on one order.

Finally, we will look at the logistics. Taxes, brokerage accounts, and the daily routine. I want this to be practical. I want you to know exactly what to do when you log in on Monday morning. I want you to spend less time staring at charts and more time enjoying the income you've generated.

A Warning and A Promise

Before we turn the page, I need to offer a warning. This approach requires more from you than index investing. You cannot just set up an automatic transfer and fall asleep for thirty years. Running a factory requires a foreman. You will need to learn a new language. You will need to make decisions. You will need to manage your emotions when prices swing wildly.

There is no free lunch in finance. The "cost" of the superior income we generate is the time we spend learning the craft and the discipline we must maintain to execute it.

But here is the promise: Once you understand how to generate your own cash flow, you will never look at the stock market the same way again.

You will stop hoping for a bull market and start profiting from *any* market. You will stop worrying about what the Federal Reserve does with interest rates and start adjusting your own rates of return. You will regain control.

The goal is not to have the highest score on the leaderboard when you die. The goal is to build a machine that funds your life, protects your family, and gives you the freedom to do what you want, when you want, without asking the stock market for permission.

Grab your hard hat. We have a factory to build.

Chapter 1

The Income Factory Philosophy and Foundations

Imagine if, instead of constantly checking stock prices and hoping they go up, you simply checked your bank account to see how much new cash arrived today. That is the shift from being a 'price speculator' to becoming an 'income manufacturer.' Before we start pulling levers and pushing buttons in the Income Factory, we need to inspect the foundation. This chapter is about rewiring your brain to value cash flow over paper gains and ensuring your personal finances are robust enough to handle the powerful tools-like options and leverage-that we will introduce later.

1.1 Shifting to a Cash-Flow Mindset

The most expensive sentence in investing is four words long: *"I'll wait until..."*

- I'll wait until the market calms down.
- I'll wait until rates stop moving.
- I'll wait until after earnings.
- I'll wait until the election.
- I'll wait until it feels safer.

That sentence quietly turns investing into a game of timing—and timing turns you into a weather forecaster who has to be right *twice*: when to get in, and when to get out. Worse, it makes your success depend on the one thing you can't control: what other people are willing to pay for the same asset tomorrow.

A cash-flow mindset flips the script. It says: *"I don't need to predict. I need to produce."* The scoreboard changes from "How high did my account balance float today?" to "How much cash did my portfolio manufacture this month?" That shift sounds subtle. In practice, it changes your behavior in every market environment—especially the ugly ones.

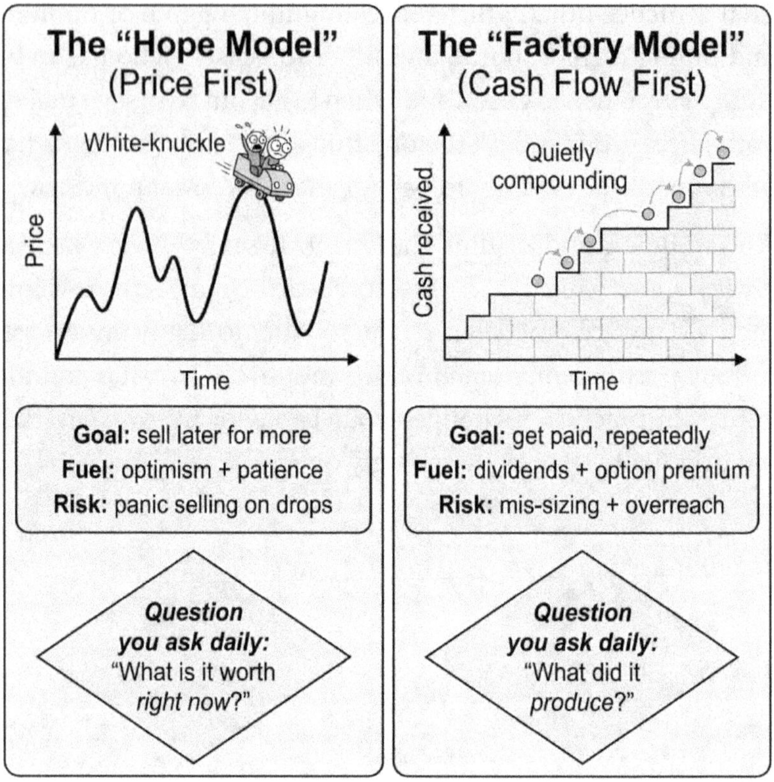

Picture two neighbors who each own the exact same house on the same street.

Neighbor A refreshes Zillow three times a day. If the estimate is up, they feel smart. If it's down, they feel vaguely ill. Their mood swings with a number no one is offering them in cash.

Neighbor B has a tenant. Every month the rent hits the bank account. They don't ignore the house value—they just don't *live* inside it. The rent is the point. The house value is background noise unless they need to refinance or sell.

Most investors are trained to behave like Neighbor A. Financial media is built for it. It has to be. "A steady stream of dividend payments arrived on schedule" is not much of a headline. "Stocks plunge 900 points!" sells ads.

The Income Factory philosophy is simply choosing to think like Neighbor B. The portfolio isn't a scoreboard; it's a machine. Your job is to keep it operating: well-capitalized, properly maintained, and productive in multiple seasons.

Why the ``buy low, sell high" story breaks your brain

"Buy low, sell high" is not wrong. It's just incomplete—and psychologically brutal.

The missing middle is *time*. You can buy low and still watch the price go lower. You can be right on value and wrong on timing. Meanwhile, the market charges you a daily emotional fee: uncertainty.

This is where good investors develop bad habits:

- *They become allergic to volatility.* Normal price movement starts to feel like danger instead of weather.
- *They confuse activity with control.* More checking, more trading, more "risk management" that is really just anxiety in a blazer.
- *They start negotiating with themselves.* "If it gets back to even, I'll sell." That's not strategy; that's bargaining.

A cash-flow mindset doesn't eliminate uncertainty—nothing does. It changes what you're listening for. Instead of waiting for the market to applaud you with a higher price, you're collecting cash along the way. The market can boo. The tenant still pays.

Cash flow is not magic (and it's definitely not ``free")

A quick reality check keeps this philosophy honest: cash flow is not a cheat code. Dividends and option premium do not appear from thin air. They are *priced*.

A dividend is a cash distribution from a company to shareholders. After a dividend is paid, the company has less cash than it did the day before. Option premium is compensation you receive for taking on specific obligations (like agreeing to buy or sell shares at a set price). If you collect premium, you are *selling insurance*. Sometimes the policyholder makes a claim.

So why emphasize cash flow at all?

Because humans are not spreadsheets. We're behavior. We're sleep. We're the decisions we make when the market is down 18% and our group chat is suddenly full of amateur macroeconomists.

Cash that arrives in your account can be reinvested, saved, used to pay bills, or simply kept as dry powder. It gives you *options* in the ordinary English sense of the word. And options—choices—are what keep investors from making forced, expensive mistakes.

The real product: staying power

The Income Factory is less about "getting rich" and more about "not getting shaken out."

In a pure price-return world, you need the market to cooperate on *your timeline*. If it doesn't, you're stuck staring at a red number and hoping your patience outlasts your fear.

In a cash-flow world, you are building something that can feed itself:

- Cash distributions can be reinvested into more income-producing assets.
- Option premium can be used to lower cost basis, diversify, or build a reserve.
- Regular inflows create a natural discipline: you focus on *process* instead of headlines.

This is what "factory" really means. A factory doesn't care if the billboard outside says the neighborhood is trendy this week. It cares about input costs, output quality, and downtime. It cares about not breaking.

A tale of two statements: ``My portfolio is down" vs. ``My income is up"

Here's the strange emotional math of investing:

> When your portfolio value drops, it feels like something was *taken* from you.
> When your income arrives, it feels like something was *given* to you.

Even when both are "just numbers," they land differently.

Say your account is worth $500,000. A normal ugly month takes it down 8% to $460,000. That hurts. If your strategy is "sell later at a higher price," your brain instantly starts doing courtroom work: *Is this temporary? Should I reduce risk? What if it's 2008 again?*

Now imagine that same month you receive $3,000 in dividends and option premium.

Is $3,000 bigger than a $40,000 drop? No.

Does it change what you *do* next? It can. Because cash flow reduces the feeling of helplessness. It gives you a lever: you can reinvest while prices are down, you can add selectively, you can pay bills without liquidating at a bad time. It doesn't eliminate drawdowns; it makes them survivable.

A portfolio that produces cash is a portfolio that can keep you from turning a temporary decline into a permanent decision.

Income vs. total return: the grown-up reconciliation

A common misunderstanding shows up right here, so let's clean it up before it turns into a religion.

Total return is the combination of:

- price change (your shares go up or down), plus
- income (dividends, interest, distributions, option premium), minus
- costs (fees, taxes, trading friction).

The Income Factory doesn't reject total return. It reframes it.

Think of total return as your *long-run report card*. Think of cash flow as your *day-to-day payroll*.

Payroll matters because it funds real life. It also funds behavior. People who get paid are less likely to flinch. People who are "waiting for appreciation" are more likely to make desperation trades when the waiting gets uncomfortable.

The aim is a portfolio where the income component is large enough to be meaningful, steady enough to be relied upon, and flexible enough to reinvest into future production. If you only chase the highest yield number you can find, you'll eventually buy something whose "income" is really just risk wearing cologne. But if you ignore income entirely, you risk becoming dependent on selling shares at whatever price the market is offering that day—which is its own form of fragility.

The rental-property test (and what it teaches you about stocks)

A rental property is a wonderfully clarifying thought experiment because it forces you to separate *cash flow* from *quoted value.*

Imagine you buy a duplex for $400,000. You put down a deposit, you have a mortgage, you collect rent, you pay expenses, and ideally you end up with positive monthly cash flow.

Now imagine someone mails you a letter every day with a different "estimated value" of the duplex:

- Monday: $410,000
- Tuesday: $395,000
- Wednesday: $420,000
- Thursday: $405,000

You would find this amusing for about five minutes and then recycle it. You'd care about whether the roof leaks, whether the tenant pays, whether your insurance is current, and whether the cash flow covers the bills. The daily estimate is not the business.

Stocks and funds are the opposite experience. The "estimate letter" is tattooed on your phone, updated by the second, and the world screams at you to interpret it as meaning.

The Income Factory approach is choosing to treat quoted prices the way a landlord treats Zestimate fluctuations: useful context, not a daily emotional referendum on your intelligence.

Where options fit the philosophy (without getting technical yet)

Options can sound like a different sport—something reserved for screens full of colored numbers and people who drink espresso like it's oxygen. But they have a surprisingly ordinary economic role:

> Options move money from people who want *certainty* to people willing to provide *insurance*.

When you sell certain types of options in a disciplined way, you are collecting insurance premiums. You're agreeing to transact at a pre-set price under certain conditions, and you get paid for taking that obligation.

That payment—the premium—shows up as cash flow.

Used responsibly, options can complement dividends by:

- generating income in flat or choppy markets,
- creating a systematic "sell high" (covered calls) and "buy lower" (cash-secured puts) framework,
- turning volatility (which usually feels like stress) into something that can pay you.

Used irresponsibly, options can create fake "yield" that collapses the moment the market makes a real move. A factory that runs too hot eventually melts a belt.

The cash-flow mindset doesn't say "always sell options." It says "if you use power tools, do it with safety guards and a clear purpose: reliable production."

The emotional dividend: why cash makes you calmer (even when prices misbehave)

There's a reason retirees love dividends, business owners love recurring revenue, and landlords love rent. Regular cash receipts do something to the human nervous system. They replace suspense with rhythm.

Rhythm changes the kinds of mistakes you make.

In a pure appreciation strategy, a market dip can trigger a series of bad questions:

- "What if I'm wrong?"
- "What if it never comes back?"
- "Should I just get out and re-enter later?"

In a cash-flow strategy, the questions are more operational:

- "Is the income stream intact?"
- "Are any positions sized too large for the risk?"
- "Is this a chance to reinvest at better prices?"

Those are still serious questions. They're just less likely to end with you rage-selling at 10:17 a.m. because you saw a chart and felt your stomach drop.

The trap: treating yield like a personality trait

A caution that saves a lot of pain: yield is not a trophy. Yield is a *price tag*.

If an investment is paying 12% when everything else similar is paying 6%, the market is not handing out gifts. It is warning you. Something is riskier, more leveraged, more cyclical, or less trusted. Sometimes that risk is worth taking in a diversified portfolio. Sometimes it's an invitation to step on a rake.

The cash-flow mindset is not "maximize yield at all costs." It's "build a stream you can live with."

"Live with" means:

- you understand where the cash comes from,
- you can tolerate the price volatility without being forced to sell,
- the strategy survives bad markets *without* requiring heroics.

Factories don't brag about how fast they can run for one hour. They care about uptime.

A practical way to measure progress: the paycheck chart

If you want this philosophy to change your decisions, you need a scoreboard that reinforces it.

Most broker apps show you what they want you to stare at: today's gain/loss. It's addictive because it's immediate. It's also a mood disorder disguised as information.

A cash-flow investor tracks something else: *trailing 12-month income*. That means adding up the cash you actually received in the last year from dividends, interest, and net option premium.

Why trailing 12 months?

- It smooths out lumpy payments.
- It captures seasonality.
- It turns the question from "What happened today?" to "What is this machine producing over time?"

When you watch that number rise—even slowly—you start to feel progress that doesn't require a bull market's permission.

But what about selling shares? (The honest answer.)

Sometimes selling shares is perfectly rational. In fact, even the most income-focused investor may sell:

- to rebalance risk,
- to exit a position whose fundamentals changed,
- to fund a large one-time expense.

The difference is *dependency*.

In a pure total-return lifestyle plan, you often depend on selling shares for spending. That can be fine when markets are kind and dangerous when they're not. If the market drops early in retirement and you must sell anyway, you risk "sequence-of-returns" damage: locking in losses while shrinking the base that could recover later.

A cash-flow approach aims to reduce that dependency by funding a meaningful portion of spending from ongoing portfolio production. You're still an investor, not a medieval peasant living off the land. You're just trying to avoid the position of having to liquidate at fire-sale prices to pay a very unglamorous bill like property taxes.

The mental shift, distilled

The traditional model trains you to think like this:

> *"I buy assets, then hope other people will pay more later."*

The Income Factory model trains you to think like this:

> *"I buy assets (and sometimes sell carefully chosen obligations) that pay me for owning them."*

One is suspense-driven. The other is production-driven.

One encourages you to check prices. The other encourages you to check processes.

One makes you feel smart when the market is generous. The other helps you behave smartly when it isn't.

Actionable takeaway: build a two-line ``income dashboard"

Open a spreadsheet (or notes app) and create a tiny dashboard you can update once a month in five minutes:

1. *Cash Received (Last 30 Days):* dividends + interest + net option premium actually credited to your account.
2. *Trailing 12-Month Cash Received:* the sum of the last 12 monthly totals.

Then do one more thing that feels almost childish but works: write a single sentence under the numbers.

> *"My portfolio's job is to produce cash. Price is weather."*

If you only get one benefit from this philosophy, let it be this: the next time the market throws a tantrum, you'll have a calmer metric to look at—a number that reflects what your investments *did*, not what strangers *think* they're worth today.

1.2 The Raw Materials: Budgeting and Financial Reality

A portfolio can survive a bad week. Your life usually can't.

That's the part people skip when they get excited about "income strategies." They picture the brokerage account as a self-contained machine: deposit money, pull levers, receive cash. In real life, the machine is bolted to the floor of your household. If the dishwasher dies, if the car needs brakes, if a medical bill shows up with the warmth of a parking ticket, the portfolio doesn't get to vote. It gets tapped.

The Income Factory only works when it's fed with money you can truly leave alone. Not "I think I can leave it alone." Not "I'll probably be fine." *Truly*. Because the quickest way to sabotage an income plan is to turn your investing account into your checking account's overdraft protection.

The raw material has a name: *investable surplus*. It's the portion of your cash flow that can be deployed into the factory without threatening your rent, your groceries, or your sleep.

The two lies we tell ourselves about money

Most budgeting advice fails because it treats the numbers like they're the problem. The numbers are usually innocent. The problem is the story we attach to them.

Lie #1: *"My income is what I make."*

Your income is what you *keep*. A household that earns $150,000 and spends $149,500 has an "income" of $500 in any practical sense. That's not a moral judgment; it's physics. Cash has to exist before it can be invested.

Lie #2: *"My expenses are fixed."*

Your expenses are sticky, not fixed. They creep. Subscriptions reproduce like rabbits. "Small treats" become a category. One "temporary" convenience becomes a permanent habit. A budget isn't a document you make once; it's a flashlight you keep turning on in corners where money likes to hide.

Here's the good news: you don't need a perfect budget. You need a *true* one.

The adult definition of ``budgeting": a cash-flow x-ray

Think of budgeting less like a diet and more like lab work.

A diet tells you what you *should* eat. Lab work tells you what's actually happening inside your body, whether you approve of it or not. Budgeting, done correctly, is an x-ray of your household cash flow: where money comes in, where it leaks out, and whether you're quietly relying on debt to patch holes.

The most useful budget is not organized by guilt ("bad spending") versus virtue ("good spending"). It's organized by *stability*:

- predictable inflows vs. lumpy inflows,
- essential outflows vs. discretionary outflows,
- monthly bills vs. ambush bills (insurance deductibles, car repairs, travel, gifts, taxes).

This classification matters because investing strategies—especially options—hate surprises. A surprise expense doesn't just cost you money. It can force you to close positions at the wrong time, turn a paper drawdown into a real loss, or push you into selling "safe" assets to keep "risky" ones alive. That's backwards triage.

Your portfolio is not a piggy bank (and not a credit card)

A lot of households run a quiet, informal system: spend normally, then use "whatever's left" for investing. That sounds sensible until you notice the hidden assumption: that there *will* be something left, and it will be left at the *right time*.

Markets don't coordinate with your calendar. The month your water heater dies is often the same month the market feels like auditioning for a disaster movie. If you're forced to pull money from your portfolio in that moment, you're not "taking profits" or "funding a goal." You're being mugged by timing.

Investable surplus is money that remains after you've paid for:

- your current lifestyle,
- your near-term obligations you can predict,
- the ones you can't predict but *know* will happen.

The factory runs on *surplus*. If you feed it money you'll need for Tuesday, it will eventually make you sell on Tuesday.

A simple way to find your investable surplus (without turning life into a spreadsheet hobby)

The goal is clarity, not perfection. Use a 90-day snapshot. Ninety days is long enough to catch reality—and short enough that you'll actually do it.

Step 1: Track cash in. Use take-home pay, not gross pay. Include side income, reimbursements that are truly income, and any recurring support.

Step 2: Track cash out. Use bank and credit card statements. Don't "estimate." Estimating is how money disguises itself.

Step 3: Normalize the ambush bills. This is the move that turns a budget from optimistic fiction into something you can build on. Make a list of irregular-but-real costs and convert them to a monthly number:

- car insurance every six months,
- annual property taxes,
- holiday spending,
- routine medical/dental,
- home maintenance (even renters have moving costs, deposits, replacing broken stuff).

Then divide by 12 and treat it like a monthly bill you pay to your future self.

Step 4: Create a buffer line. Not an emergency fund yet—just a monthly buffer so you don't run the household on fumes. This is the "oops" money that keeps you from swiping a credit card for a minor surprise.

Step 5: What's left is your investable surplus. That's the fuel you can shovel into the factory without quietly setting fire to the rest of the building.

If you like formulas, here's the whole thing in one line:

$$\text{Investable Surplus} = \text{Net Income} - (\text{Essential Expenses} + \text{Normalized Irregulars} + \text{Monthly Buffer})$$

You'll notice what's missing: "market opportunity." That's intentional. A sustainable investing plan does not depend on you being in the mood, the market being cooperative, or your boss not being weird this quarter.

The $400 test: why ``small" emergencies aren't small

People talk about emergencies like they're rare meteor strikes. In most households, emergencies are more like potholes. You don't plan your route around them. You just hit them.

The Federal Reserve has a famous survey question about whether adults could cover a $400 emergency expense using cash or its equivalent. The reason this question stings is not that $400 is huge. It's that $400 is *common*. It's tires. It's a copay plus a prescription. It's a broken phone when you need it for work. It's the bill that shows up right when you were about to feel "caught up."

If $400 would force you onto a credit card, your investable surplus is not investable yet. Not because you're "bad with money," but because you're trying to run a factory without inventory. Every time a pothole appears, you'll be tempted to yank money out of the market at the worst moment, or you'll drift into debt and pay interest so your portfolio can keep earning dividends. That's like paying a fee to keep a coupon.

The first job of budgeting is to make sure small shocks don't create big decisions.

Surplus is not what you save; surplus is what you can *repeat*

A powerful way to think about investable surplus is this: it must be *repeatable*. A one-time windfall can be invested, sure. But it doesn't tell you whether your factory can keep running.

A repeatable surplus has two traits:

- It doesn't require heroics (no "I'll just not spend anything for the next two months").
- It doesn't cannibalize essentials (no skipping insurance, ignoring maintenance, or floating necessities on credit).

If your surplus only exists when you don't go to the dentist, it's not surplus. It's deferred pain with interest.

If your surplus only exists when nothing breaks, it's not surplus. It's luck.

The goal is a number you can hit even in an ordinary, slightly annoying month—because ordinary, slightly annoying months are the ones you'll get most often.

The hidden enemy of income strategies: the household ``margin call"

Brokerage margin calls are dramatic. They have emails. They have deadlines. They have that special tone of "fix this now."

Household margin calls are quieter. They look like:

- the checking account dipping below zero,
- a credit card balance that stops going back to zero,
- moving money around "just for a week" that somehow becomes permanent,
- borrowing from your tax withholding because "I'll deal with it later."

When your household is living on a thin margin, any investing strategy becomes riskier than it needs to be. Not because the strategy changed, but because your *ability to hold it* changed.

Options and dividends can generate cash flow, but they can't fix a structural deficit. If the household is short $600 a month, a portfolio that pays $600 a month is not "income." It's a patch. And patches have a way of coming off in the rain.

Budgeting that doesn't feel like punishment: design, not willpower

Willpower budgeting is the one where you promise yourself you'll "be good" and then feel personally attacked by a latte.

Design budgeting is the one where you make the default behavior smart, so you don't have to negotiate with yourself every day.

Three design moves are especially effective because they attack friction, not character:

- *Automate the first step.* Have a set amount move automatically on payday into a separate "buffer/surplus" account. Automation turns saving from a decision into a system.
- *Create separation.* Money that sits in the same checking account as your bills gets mentally spent twice: once when you see it, and again when you actually spend it.
- *Use a "boring" account for stability.* A high-yield savings account or money market fund isn't exciting, which is exactly why it works as a staging area. Exciting accounts invite tinkering.

Notice what's not on the list: shame. Shame is a terrible financial planner. It makes people hide from their numbers, which is the opposite of what you need.

A quick reality check on ``making income" from investments

If your investable surplus is currently $0, that's not the end of the story. It's the beginning of a better one.

Here's the honest sequence:

- A budget creates surplus.
- Surplus funds investments.
- Investments create cash flow.
- Cash flow can accelerate surplus (reinvested income grows the machine).

Flip that sequence and you get the fantasy version:

> *"I'll invest first, and the income will fix the rest."*

Sometimes people get away with that when markets are roaring. When markets are ordinary (or rude), it tends to end with a forced sale, a tax surprise, or a credit card balance that becomes a roommate.

The factory is an *accelerator*. It is not a substitute for positive household cash flow.

The most practical tool: a ``minimum viable budget"

If the word "budget" makes you want to close the book and go reorganize your sock drawer, use a minimum viable budget. It has exactly two numbers and a rule.

Number 1: Essential monthly nut. This is what it costs to keep life running: housing, utilities, basic groceries, transportation for work, insurance, minimum debt payments, child care you can't opt out of.

Number 2: True monthly net income. Again: after tax, after deductions, after reality.

The rule: If the gap between those numbers is small or negative, investing becomes a fragile hobby until you fix the gap. If the gap is comfortably positive, you can build reserves and invest the surplus with far less risk of being forced into a bad move.

This approach is blunt on purpose. Clarity is kinder than complexity.

Turning surplus into fuel: staging and pacing

Once you've found surplus, the next temptation is to shove it into the market immediately. Sometimes that's fine. Often, it's smarter to stage it.

A staged approach looks like this:

- First, build a small cash buffer that prevents household margin calls.
- Then, move surplus into the investing account on a schedule (weekly or monthly).
- Keep contributions steady even when markets are noisy, because your contribution schedule should be based on your paycheck, not your feelings.

Pacing matters because it protects you from the emotional whiplash of "I invested it all and the market dropped the next day." That experience turns adults into superstitious creatures.

A paced plan keeps you investing through weather without needing to be brave every time.

Actionable takeaway: calculate your investable surplus in 30 minutes

Set a timer for 30 minutes and do this with whatever tools you already use (bank app, credit card app, notes):

1. Write down last month's net income (what hit your bank account).
2. Write down last month's total spending (credit cards + bank withdrawals).
3. List your irregular-but-real expenses for the year (insurance, taxes, car/home, holidays). Divide the total by 12 and write that monthly number down.
4. Subtract spending and the monthly irregular number from net income.

If the result is positive, that's your *first draft* investable surplus. Start feeding the factory with a conservative portion of it (not all of it), and keep refining as you notice what you missed.

If the result is negative, you just achieved something valuable: you replaced vague anxiety with a measurable target. Your job becomes simple and concrete—increase income, reduce spending, or both—until the number flips. Once it flips, you'll be investing with money you can actually afford to leave alone, which is the difference between an income plan that feels empowering and one that feels like a tightrope.

1.3 Building the Moat: Liquidity and Emergency Reserves

The market has impeccable timing. It loves to drop the week your transmission dies.

That isn't superstition; it's simply how stress feels when two unrelated problems collide. A portfolio drawdown is tolerable when your life is calm. A surprise bill is tolerable when the market is calm. Stack them together and you get the most dangerous force in personal finance: *the forced sale*. You don't sell because the plan changed. You sell because the rent is due and the account happens to be red.

Liquidity—plain, boring, unglamorous liquidity—is what stops that chain reaction. It's the moat around the castle. It doesn't make you richer in a bull market. It makes you harder to defeat in a bad one.

And if you want to use options for income, liquidity isn't optional. Options are contracts. Contracts come with obligations. Obligations plus bad timing plus a thin checking account is how people end up "learning a lot" in the most expensive way possible.

Liquidity is not the same thing as wealth

A person can be wealthy and illiquid, and the reverse can be true in smaller ways.

The classic illiquid-rich story is someone who owns a valuable business or a pricey house and still panics over a $2,000 bill because the money isn't sitting anywhere spendable. But you don't need a seven-figure net worth to experience the same problem. It happens every time your financial life is tied up in things that are hard to convert to cash quickly without damage: investments in a drawdown, a car you need to get to work, a retirement account you don't want to touch, or even "cash" that's technically available but would trigger taxes or penalties.

Liquidity means *spendable money with low drama*. Not "I can probably sell something." Not "I can float it on a credit card." Not "my brokerage app says I have buying power." Actual cash you can deploy when life gets rude.

That's the moat. It's not about maximizing returns. It's about avoiding the one outcome that permanently impairs returns: selling good assets at bad prices because you had no choice.

The real risk isn't volatility; it's fragility

Volatility is the market's mood swings. Fragility is your inability to tolerate them.

A portfolio can be perfectly designed on paper and still fail in reality if the investor has no buffer. When people say "I can handle risk," what they often mean is "I can handle risk *when nothing else goes wrong*."

Liquidity changes the environment in which risk happens. It turns a market drop from an emergency into an inconvenience. It buys you time—and time is what investing needs to work.

The cruel truth is that many investors don't lose money because they picked the wrong assets. They lose money because they picked the right assets with the wrong cash position. They were right, but they were early, and they couldn't afford to wait.

The $400 pothole and what it reveals

A $400 surprise expense is not a financial earthquake. It's a pothole. It's what happens when your kid breaks their glasses, when your tire finds a nail, when the dentist says "we should take care of this now."

The reason that $400 question (could you cover it with cash or its equivalent?) is such a powerful diagnostic is that it tests liquidity, not income. Plenty of households earn decent money and still fail the test because every dollar already has a job, and the jobs don't include "unexpected."

If $400 would push you onto debt, you don't have an investing problem. You have a moat problem. Without liquidity, any investing strategy becomes a high-wire act. With liquidity, even a plain strategy becomes more powerful because it can be held through weather.

What an emergency fund actually does (besides ``being there")

People describe emergency funds like they're a stash of cash you hope never to use. That framing makes them feel like dead weight.

A better framing: an emergency fund is a *behavioral tool*. It prevents you from making dumb, permanent decisions during temporary pain.

In a market downturn, an emergency fund does three critical jobs:

- It keeps your bill-paying system separate from your investing system, so you don't liquidate investments to handle routine chaos.
- It gives you the courage to stay invested when headlines are designed to scare you out.
- It protects your income strategy from being cannibalized at the worst moment. (Nothing kills "compounding" like pulling the plug mid-cycle.)

That last point matters for options. If you sell options for income, you're accepting that the market might hand you shares (via puts) or cap your upside (via calls) during volatile stretches. Those periods are exactly when you want to be able to manage positions patiently, not urgently. Liquidity turns position management from a panic response into a calm decision.

The 3--6 month rule is a starting line, not a finish line

A common rule of thumb is an emergency fund of roughly 3–6 months of living expenses. It's popular because it's simple and usually sane.

But the right number isn't just about months. It's about *how your life actually breaks*. A single person with stable employment and low fixed costs may be comfortable at the low end. A household with irregular income, a single earner, variable medical costs, or high fixed obligations might need more. The point isn't to worship a number. The point is to make sure a normal run of bad luck doesn't force you to torch your investing plan.

Also: "living expenses" should mean *essential* living expenses, not "everything we spend when life is good." Emergency funds are about keeping the lights on, not keeping brunch intact.

Liquidity tiers: cash, near-cash, and long-term capital

Most people treat liquidity as binary: either you have cash or you don't. In practice, smart households build *tiers*—a layered system that matches money to time.

Think in three layers:

Tier 1: Immediate cash (the shock absorber)

This is money you can access today or tomorrow without selling anything risky. It covers the potholes: small emergencies, temporary cash flow timing issues, minor repairs. The job of Tier 1 is to prevent you from reaching for debt for every surprise.

Tier 1 should be *boring and accessible*. A checking account buffer or a high-yield savings account works. The yield is not the point. The speed is.

Tier 2: Short-term reserves (the bridge)

This is where "3–6 months" usually lives. It's there for the bigger disruptions: job loss, health event, moving costs, family emergency travel, a major repair. It's not meant to swing around with the market. It's meant to sit there, ready, so you don't have to sell stocks the week they're down 25%.

Tier 2 can still be simple and liquid: savings, money market funds, short-term Treasury funds, ladders of CDs if you understand the access terms. The principle is the same: it should be spendable without having to ask the market's permission.

Tier 3: Long-term capital (the factory floor)

This is the money you invest for decades: dividend stocks, bond funds, options-based income strategies, whatever fits your plan. This tier is allowed to be volatile because it's protected by the first two tiers. It's the engine, not the airbag.

This tiered structure is how you keep your investing strategy from being hijacked by life. You stop treating the portfolio as the emergency fund and treat the emergency fund as the portfolio's insurance policy.

Why options trading makes the moat more important, not less

Options trading can produce cash flow. That's attractive for obvious reasons. But options also introduce a kind of *timing risk* that traditional buy-and-hold investors don't experience as sharply.

Here's the problem in plain language: options have *deadlines*. Contracts expire. If you sell a put and the stock drops hard, you might get assigned shares at a price that looked reasonable a month ago and feels spicy today. If you sell a call and the stock rips higher, you might have to choose between letting shares go or adjusting the position.

In both cases, you want to be able to make a calm choice:

- Do I have cash to take assignment if it happens?
- If I need to roll (extend) a contract, do I have the flexibility to do it without dipping into bill money?
- If the market is stressed and spreads are wider, can I wait for a better fill?

Liquidity is what turns these from scary questions into ordinary ones. Without liquidity, every option position is secretly connected to your rent payment. That's when "income" strategies become stress strategies.

A quick checklist: signs your moat is too small

A good moat is quiet. A small moat announces itself.

Watch for these signals:

- You keep a credit card balance because "the market will bounce back soon."
- You feel tempted to stop investing every time you have an unexpected bill.
- You find yourself thinking, "I can't afford a downturn right now."
- You're using a brokerage margin loan (or even just the idea of it) to smooth household cash flow.

- You're selling positions to pay for predictable expenses (insurance, taxes) because you didn't stage cash for them.

None of these make you a bad investor. They simply mean your household is running with too little slack. The fix is not necessarily to abandon investing; it's to widen the moat until investing becomes stable.

Liquidity doesn't have to be huge to be powerful

There's a seductive all-or-nothing mindset that shows up with emergency funds: "If I can't save six months of expenses, why bother?"

Because a small buffer changes your trajectory immediately.

Even a few hundred dollars can break the habit of turning every surprise into debt. A few thousand can stop the bleeding from high-interest interest charges. A month of expenses can prevent a temporary disruption from becoming a financial spiral. Liquidity grows like any other asset: it starts small, then becomes self-reinforcing because it reduces crisis behavior.

A moat doesn't need to be a mile wide to matter. It just needs to be wide enough that the first arrow doesn't hit the wall.

Where to keep the moat (and where not to)

The best place for emergency money is where it won't be tempted into adventure.

Good traits:

- easily accessible,
- low risk of principal loss,
- low chance you'll "optimize" it into something complicated,
- psychologically separate from your investing account.

A common mistake is holding emergency reserves in the same brokerage account you use to trade. Even if the money is in a money market fund, it's sitting next to buttons that whisper, "You could put this to work." In a frothy market, those whispers get loud.

Separation is underrated. A moat works better when it's literally in a different place.

Liquidity and the investor's superpower: not being forced

A strange thing happens when you build real liquidity: you become harder to manipulate.

Markets manipulate through urgency. Financial media manipulates through urgency. Many bad financial products sell urgency. "Act now." "Limited time." "Don't miss out."

Liquidity makes you the person who can say, "No thanks. I'll decide on my schedule."

That's not just emotionally pleasant; it's profitable. The ability to wait is an edge. The ability to wait is also what allows an income strategy to behave like a factory instead of a casino.

Actionable takeaway: build your moat in two numbers and one automatic transfer

Set up a moat with a simple, concrete plan you can start this week:

1. Calculate your *Essential Monthly Spend*: housing, utilities, basic food, transportation, insurance, minimum debt payments.
2. Choose two targets:
   - *Tier 1 target:* $500–$2,000 (enough to make potholes boring).
   - *Tier 2 target:* 3 months of essential spending (a solid first bridge).
3. Set an automatic transfer on payday into a separate savings account labeled *Moat*. Start with an amount that won't cause a backlash in your checking account—even $25 or $50 per paycheck is legitimate.

Then add one rule that keeps the moat doing its job: *If the moat gets used, the next surplus dollars refill it before new investing contributions increase.*

That single rule is how you keep "one surprise bill" from turning into "a permanent fragile system." It's also how you earn the most valuable benefit an income investor can have: the freedom to hold, adjust, and reinvest through ugly markets without being forced to sell at the worst possible time.

1.4 Respecting the Power Tools: Debt and Leverage

Fire is a wonderful invention right up until it's in the curtains.

Leverage works the same way. Used with care, it heats the home: it can improve efficiency, smooth cash flow, and help a well-run plan do more with the same starting capital. Used casually, it doesn't just *increase risk.* It changes the *shape* of risk. It turns ordinary market weather into something that can break the account—quickly, and sometimes permanently.

The Income Factory mindset is about building a machine that keeps producing. Leverage is the part of the machine that can spin faster than it should. If you don't install guardrails, it won't merely wobble. It can throw a belt.

Leverage, in plain English

Leverage is using borrowed money (or borrowed exposure) to control more assets than your cash would normally allow.

If you have $10,000 and buy $10,000 worth of stock, you're unleveraged. If you have $10,000 and borrow another $10,000 to buy $20,000 worth of stock, you're roughly at *2x leverage*. You now own twice the market exposure, which means your gains can be bigger. Your losses can be bigger too.

The key detail most people miss: leverage doesn't just amplify outcomes. It compresses time. You can reach the "I can't recover from this" zone much sooner.

That's why leverage has a reputation for being a villain in stories about blow-ups. It's not because leverage is inherently evil.

It's because it's impatient. It punishes sloppy position sizing with unusual speed.

Margin: the most common form of leverage investors touch

In brokerage language, *margin* is a loan from your broker that lets you buy (or hold) more securities than your cash alone would allow. Your securities serve as collateral. The broker is not doing this out of kindness. They charge interest, and they require you to maintain enough equity in the account to protect their loan.

Margin is a power tool because it's frictionless. It doesn't feel like a loan application. There's no banker asking whether your plan makes sense. It's a toggle, a setting, a quiet feature that can sit unused for months until the day it becomes the easiest button to press.

On that day, margin can feel like relief: *"I don't have to sell; I'll just borrow."* Sometimes that's a reasonable bridge. Sometimes it's the financial version of picking up a rattlesnake because you don't want to walk around it.

The invisible bill: margin interest

Margin loans come with interest, and the meter runs whether your investments are up, down, or starring in a tragic documentary.

That interest is not just a cost; it's a behavioral trap. It can encourage you to "stay in" a position longer than you should because you don't want to realize a loss. Meanwhile, you're paying to avoid making a decision. That's not investing; that's renting denial.

For an income-focused investor, margin interest has a particularly sharp edge: it competes directly with your yield. If you're earning 6–8% in portfolio income but paying a meaningful borrowing rate, the spread can shrink fast. The strategy may still work, but now it has a new dependency: financing conditions. You've added a second weather system.

The margin call: the broker's emergency exit (that you don't control)

A *margin call* happens when the equity in your margin account falls below the required level. This can happen because your holdings drop in value, because requirements change, or because concentration increases the broker's risk.

When a margin call happens, you don't get to negotiate with the market on your timeline. The broker can require you to add cash or securities quickly. If you don't, they can sell assets to bring the account back into compliance.

The truly unsettling part is not the idea of being asked to add money. It's the loss of control:

- positions may be liquidated without warning,
- you may not get to choose *which* positions are sold,
- the liquidation may happen in a fast market, at bad prices.

That's the opposite of the Income Factory ethos. A factory survives by staying operational. A margin call is downtime enforced from the outside.

The leverage seesaw: why 2x is more than ``twice as risky"

People hear "2x leverage" and mentally translate it as "twice the volatility." That's directionally correct, but it understates the real danger.

With no leverage, a 50% decline is awful but not necessarily fatal if you can hold. With 2x leverage, a 50% decline can be catastrophic because your equity can approach zero long before the underlying asset does.

The math is simple and merciless:

- Unleveraged: $100 becomes $50 after a 50% decline.
- 2x leveraged (conceptually): $100 of equity controls $200 of assets. A 50% decline in the assets is a $100 loss, which can wipe out the equity.

Real margin rules often force action before "zero" because brokers have maintenance requirements. That means the practical wipeout line is usually *above* a 50% decline. You can be forced out before the market has any chance to recover.

Leverage inside options: the quiet kind

Even if you never borrow a dollar on margin, options can introduce leverage-like exposure.

An option contract controls 100 shares of stock. That's the "point" of options, but it also means small price moves in the underlying can translate into large percentage moves in the option's value. If you buy options, you can lose 100% of the premium. If you sell certain options, you can take on obligations that are much larger than the cash you collected.

Some examples of where leverage sneaks in:

- Selling puts can create exposure similar to being long shares (with downside risk) even if you only received a modest premium.
- Spreads and multi-leg trades can create concentrated exposure if you size them by "premium received" instead of by worst-case loss.
- "Naked" option selling can create losses far larger than the credit you collected, and uncovered calls can be theoretically unlimited in loss if the stock rises sharply.

Leverage isn't only borrowing. It's any structure where a small move can do a big thing to you.

Why leverage is so tempting for income investors

Income strategies have a particular psychological itch: the desire for a bigger monthly number.

You can see it in the way people talk about yield. A portfolio yielding 4% can feel "slow." A portfolio yielding 10% can feel like competence. And leverage offers a shortcut: borrow to buy more income-producing assets, or structure trades that generate more premium per dollar.

The temptation is understandable because the goal is concrete: cash flow you can see. But leverage can turn that concrete goal into tunnel vision. You start optimizing the output meter and ignore the temperature gauge.

A factory doesn't measure success by how loudly the machines hum. It measures success by how many months it stays open.

The safety rules: how grown-ups use leverage (if they use it at all)

Leverage can be a tool. The safety rules are what make it a tool rather than a gamble. The following principles are less about cleverness and more about survival.

Rule 1: Never use margin to cover lifestyle spending

Borrowing against your portfolio to pay bills is where "temporary" becomes "structural." If you need margin to make the month work, the portfolio is not supporting you; it's being harvested.

This is the investing equivalent of using a payday loan to fund a side hustle. You might get away with it, but the odds are not in your favor.

Rule 2: Size positions by worst-case discomfort, not best-case yield

Income strategies can be deceptively calming in good times. Premium arrives, dividends hit, and everything looks steady—right until the market sells off and correlations go to one.

Sizing by worst-case discomfort means asking: *If this goes against me sharply, can I hold, manage, or exit without touching my moat?* If the answer is "only if nothing else goes wrong," the position is too big.

Rule 3: Treat borrowing capacity as emergency capacity, not investing capacity

Many investors see available margin and think, "That's money I can invest." It's closer to "That's a tool I might need to prevent forced selling." Even then, it's a sharp tool.

If you decide to use margin at all, one conservative mindset is to keep it as a reserve you could tap briefly during settlement timing issues or extraordinary opportunities, rather than as a permanent way of running the portfolio.

Rule 4: Keep your liquidity moat separate from margin decisions

Your emergency reserves exist to prevent forced actions. Margin creates forced actions. Mixing them mentally is how people rationalize danger: *"I have plenty of liquidity because I can borrow."*

Borrowing is not liquidity. Borrowing is a promise.

Rule 5: Understand the tax and dividend wrinkles before layering options

Dividend income can be especially attractive when it qualifies for favorable tax rates. But some option overlays can affect holding periods or reduce eligibility if they substantially reduce risk of loss. You don't need to memorize tax code to be cautious; you need to recognize that "dividends + options" has interactions.

The practical takeaway: if dividends are a meaningful part of your plan, don't assume every options overlay is tax-neutral. Ask questions early, and keep strategies simple until you're confident.

A vivid example: the ``helpful" loan that becomes the boss

Imagine two investors with $100,000.

Investor A buys $100,000 of a diversified income portfolio. They collect income, reinvest some, spend some, and live with price swings. In a 25% drawdown, their account shows $75,000. It's unpleasant, but they still own the same shares. If their cash reserves are solid, they can keep holding and keep collecting.

Investor B borrows to boost income. They buy $200,000 of the same assets using $100,000 equity and $100,000 borrowed. In a 25% drawdown, the assets fall to $150,000. The loan is still $100,000. Their equity is now $50,000. The drawdown isn't 25%; it's 50%. At the same time, margin requirements may tighten, and the broker may demand action.

Same market. Same assets. Completely different experience.

Leverage doesn't make you wrong. It makes you less resilient.

The core principle: leverage reduces your right to be patient

Patience is the superpower of long-term investing. Leverage is the thing that can revoke it.

Without leverage, you can often endure volatility as long as you have time and liquidity. With leverage, the market can force your hand, and the forcing tends to happen when prices are worst and emotions are hottest.

That's why leverage belongs in the "power tool" category. You don't casually use a table saw when you're tired, distracted, or rushing. You use it when your stance is stable and your hands are clear.

Actionable takeaway: write your personal leverage policy before you need it

Before you ever borrow on margin or size options aggressively, write a one-paragraph policy and stick it somewhere visible. Keep it simple enough that you'll follow it when you're stressed.

Here's a template you can copy and customize:

> *"I will not use margin to pay for lifestyle expenses. I will only consider margin or leverage-like option exposure when my emergency reserves are fully funded. I will size positions so that a severe market drop would not force me to liquidate. If I borrow at all, I will cap borrowing*

at _____ of account equity, and I will reduce leverage (not add) when volatility rises."

Then do one concrete check in your brokerage settings: verify whether margin is enabled, what your account type is, and what "available to trade" really means. Half the disasters involving leverage begin with a sentence that starts, *"I didn't realize..."*

Leverage can help a factory run faster. Liquidity and discipline are what keep it from exploding when someone bumps the power switch.

1.5 The Invisible Thief: Inflation and Real Returns

Inflation is the only bill you pay that doesn't arrive in the mail.

No invoice. No due date. No customer service number. It just quietly changes the price tags while you're living your life, like someone turning the thermostat down one degree per week and waiting to see when you notice.

And the cruel part is that you *do* notice—just not all at once. You notice at the grocery store when the "normal" total becomes the "wow, already?" total. You notice when your auto insurance renewal feels like it developed a personality. You notice when your favorite lunch spot turns into a special-occasion restaurant without anyone holding a ceremony.

For an income investor, inflation is personal. It doesn't care that your portfolio "paid you." It only cares what those dollars can buy.

Nominal vs. real: the two kinds of return that get confused at parties

Most investing results are quoted in *nominal* terms: the raw percentage or raw dollars, not adjusted for inflation. If your portfolio yields 7% and inflation is 3%, it can be tempting to think, "Great—I'm up 7%."

But your lifestyle doesn't run on percentages. It runs on purchasing power.

A *real* return is what you earn *after* adjusting for inflation. In a simplified approximation:

$$\textit{Real Return} \approx \textit{Nominal Return} - \textit{Inflation}$$

So the 7% yield with 3% inflation isn't "up 7%." It's closer to "up 4% in purchasing power" before taxes and costs. Real return is the difference between feeling wealthy and merely watching bigger numbers scroll by.

This is the invisible thief's favorite trick: letting you feel progress while it quietly raises the price of everything you want.

Why inflation is extra dangerous for income strategies

Inflation hurts everyone, but it has a special way of bullying people who are focused on cash flow.

Here's why: income feels *finished*. It feels like a product you can consume.

You receive $2,500 in dividends and option premium this month. Great. But if your monthly essentials used to cost

$2,500 and now cost $2,900, the same portfolio "income" has effectively shrunk. Your factory is producing the same number of units, but the units are smaller.

Even worse, income investing can encourage a subtle complacency: if the cash keeps arriving, you assume the system is working. Inflation is the force that can make a system look stable while it's slowly falling behind.

That's why the Income Factory isn't just an income engine. It needs a growth engine too—not necessarily growth in price every day, but growth in *production capacity*.

Inflation is not one number (and your household has its own version)

Headline inflation numbers try to summarize a vast, messy economy into a single statistic. That's useful—but it's not your life.

Your personal inflation rate depends on what you buy:

- If you drive a lot, you experience fuel and maintenance inflation more intensely.
- If housing is a large portion of your budget, shelter costs dominate your reality.
- If you have recurring medical expenses, health care pricing is not an academic topic.

Two neighbors can live in the same city and have different "inflation" experiences because their spending patterns differ. One's biggest expense might be daycare. The other's might be travel. The official inflation gauge is an average; your household is a specific recipe.

An investor who ignores that specificity risks building a factory that produces cash flow in a currency that's shrinking faster than they realize.

The treadmill problem: why ``high yield" can still leave you tired

Imagine you get a raise every year, but prices rise a little faster. Your paycheck goes up. Your life doesn't get easier. You're moving, but you're not advancing.

That's the treadmill problem.

A portfolio can do the same thing:

- It pays a high nominal yield.
- You spend most of it.
- The cost of living rises.
- Your "income" buys less each year.

This is how people end up saying, "I'm making more than ever, but I feel behind." It's not a mystery. It's arithmetic.

The fix isn't "chase even higher yield." That's often how investors step into risk they don't understand. The fix is to build a system where income can *grow*.

The three ways an income factory can outpace inflation

You don't need a complicated solution. You need at least one reliable source of upward pressure.

Here are the main ways an income-focused portfolio can fight inflation:

1) Dividend growth (the classy way)

Some companies raise their dividends over time. Not every year, not forever, and not without occasional cuts—but many businesses have a culture of returning more cash as earnings and cash flow grow.

Dividend growth is powerful because it can turn a static income stream into a rising one without you doing anything dramatic. It's the portfolio equivalent of a tenant whose rent increases gradually rather than staying frozen for a decade.

2) Reinvestment (the mechanical way)

Even if the yield stays the same, reinvesting part of the income buys more income-producing assets. More assets produce more cash. That can create a compounding effect that inflation has to outrun.

This is where the factory metaphor becomes literal: if you take some of the output and buy more machines, output grows.

The mistake is thinking reinvestment is only for young investors. Reinvestment is for anyone whose future spending power matters—which is everyone who expects to be alive in ten years.

3) Selective use of options premium (the tactical way)

Options premium can add cash flow in sideways markets, but it's not automatically inflation protection. It's simply another stream of nominal dollars.

To make it inflation-resistant, the premium needs to be managed with discipline:

- don't treat premium as "free yield,"
- reserve some of it for drawdowns and opportunity,
- reinvest a portion systematically rather than spending it all.

Options can increase production, but they also increase the temptation to consume production. Inflation punishes that temptation over time.

The uncomfortable truth: spending every dollar of income is a choice

This is where income investing becomes emotionally tricky.

The whole point of the strategy is that cash arrives. It feels natural to spend it. It feels like "living off the portfolio." And for many people, that's a real goal.

But spending 100% of portfolio income is effectively saying: *"I believe my income stream will keep up with my future costs without me feeding it."* Sometimes that's true. Often it isn't.

Even small reinvestment rates matter. Reinvesting 10–20% of cash flow can be the difference between:

- a portfolio that slowly loses purchasing power, and
- a portfolio that grows its paycheck.

Inflation doesn't require you to be wrong all at once. It only requires you to be slightly underprepared for a long time.

A story you can feel: the ``same" $50,000 income that shrinks

Suppose your portfolio produces $50,000 a year in cash flow. That sounds like a milestone. For many households, it is.

Now add a fairly ordinary inflation rate over a decade. If prices rise steadily, that $50,000 buys less each year. In year one, it might cover the basics comfortably. In year ten, it might cover the basics tightly. The portfolio didn't "fail" in nominal terms. Your lifestyle changed underneath it.

That's why "I want $X per year in income" is only half a goal. The full goal is: *"I want $X per year in income that grows with costs."*

The distinction sounds pedantic until you live through it. Then it becomes obvious.

The good kind of paranoia: measuring income in purchasing power

A simple way to keep inflation from sneaking up on you is to measure your financial life in two parallel units:

- dollars received (nominal),
- dollars required to live (also nominal, but updated regularly).

The trick is to keep the "required to live" number fresh. Many households keep an old mental model of their expenses: "We spend about $4,000 a month." Then they're shocked when they look at statements and it's $5,200.

Inflation isn't just price hikes; it's also lifestyle drift that gets disguised as "normal."

If you update your essential spending number every six months, you can see whether your income factory is outrunning your cost treadmill or falling behind.

Inflation, leverage, and the double-edged sword

Inflation can make leverage look smarter than it is.

When money loses value over time, borrowing can feel attractive: you repay with "cheaper dollars." That logic is sometimes valid in certain contexts (like fixed-rate mortgages). In investing, it can become a trap, because market volatility doesn't politely align with your repayment schedule.

If you borrow to increase income, inflation becomes a justification: *"I need higher cash flow because everything costs more."* But leverage adds the risk of forced selling, and forced selling is how inflation and leverage team up to do maximum damage. Inflation raises your living costs. Leverage reduces your ability to wait. Together, they can turn a normal downturn into an account-level emergency.

The safe approach is to beat inflation through *production growth*—dividend growth and reinvestment—before you even consider beating it with borrowed horsepower.

The most useful mental model: your portfolio has a ``wage," and it needs raises

Think of your portfolio income as a wage. A wage that never gets a raise is a wage that eventually feels insulting, no matter how proud you were when you first earned it.

A healthy income factory aims for "raises" through:

- underlying businesses increasing payouts over time,
- reinvestment increasing the number of income-producing units you own,
- opportunistic buying during downturns when income assets go on sale.

The third one is the sneaky advantage of a cash-flow mindset: when you're not emotionally dependent on prices being high, you can buy when prices are low. Inflation is still there, but you're not letting it push you into desperation.

Actionable takeaway: build an inflation ``raise" into your income plan

Do one practical thing today that makes inflation less powerful tomorrow:

1. Calculate your *essential monthly spending* (the number you'd defend in a lean year).
2. Decide on an *annual raise target* for your portfolio income: start with 3% if you want a simple placeholder, and adjust based on your household reality.

3. Choose a *default reinvestment rate* of your portfolio cash flow (even 10% works) until your trailing 12-month income is rising at or above your raise target.

Then write the rule in plain language and keep it near your investing notes:

I don't just want income. I want income that gets raises.

Inflation is invisible because it doesn't feel like a crisis. It feels like "life." The investor who builds automatic raises into their income factory stops treating inflation as a surprise enemy and starts treating it like weather: always present, sometimes annoying, and never allowed to decide whether the factory keeps running.

Chapter 2

Setting Up the Factory Floor

Before we flip the switch on our income assembly line, we need to inspect the machinery. You wouldn't start a manufacturing business without knowing the difference between a drill press and a conveyor belt, nor would you sign a lease without checking the wiring. In this chapter, we are organizing your workspace. We will examine the raw materials of wealth-stocks, bonds, and funds-and learn how to spot the difference between a durable asset and a lemon. Finally, we'll set up the logistical backbone of your operation: the brokerage accounts that handle your inventory and the market rules that determine the price you pay.

2.1 The Toolbox: Stocks, Funds, and Bonds

Picture two people who both say they're "income investors."

One owns a boring utility stock that sends a dividend every quarter like clockwork. The other owns a bond fund that seems to pay every month, until one year it drops hard and the "safe income" suddenly looks less safe. Both were chasing cash flow. Both were using legitimate tools. But they were standing in different parts of the financial factory—one on the ownership floor, one in the lending bay—and the machines behave differently when the lights flicker.

Before options ever enter the picture, getting income right starts with naming the three most important tools on the workbench: stocks (ownership), bonds (lending), and funds (prepackaged baskets). The trick isn't memorizing definitions. The trick is understanding what each tool *pays you for*, and why that payment can be relied on—or cut, or delayed, or clawed back by price moves—at exactly the wrong time.

Two ways to get paid: own things or lend money

Nearly every investment you'll ever touch is a variation on one of two economic activities:

- *Ownership:* You own a slice of a business (stocks). Your claim is "what's left" after everyone else is paid.
- *Lending:* You lend money and get paid for the use of your capital (bonds). Your claim is contractual—interest first, principal later—assuming the borrower survives.

That difference—residual claim versus contractual claim—isn't academic. It's the difference between income that is *declared* (dividends), and income that is *owed* (interest). It's also the difference between collateral that can balloon in value during good times (stocks) and collateral that can behave more predictably—until interest rates move against you (bonds).

Income investors often talk as if "yield is yield," like it's a coupon you clip. In reality, each tool comes with its own built-in bargain. Stocks offer participation in growth but give no legal promise of payments. Bonds offer a legal promise of payments but limit your upside and react mechanically to interest rates. Funds offer convenience but bundle hidden decisions: what you own, how it's rebalanced, and how easily you can get in or out at a fair price.

Once you see these as different machines with different failure modes, you stop being surprised by things that regularly surprise the market.

Stocks: owning a cash-flow engine (with no guarantees)

A share of stock is a claim on a real operating business—its products, its pricing power, its competitors, its mistakes, its brand, its debt, its lawsuits, its talent, its management ego, and its ability to turn all of that into cash.

When a company pays a dividend, it's not paying you because it has to. It's paying you because it *can* and *chooses* to. That's both the beauty and the danger.

Dividends are policy, not physics. Bond interest is like rent due under a lease. Dividends are like a discretionary bonus. Boards set dividend policy based on cash flow, investment opportunities, debt covenants, and—let's be honest—what they think the shareholder base expects. Many excellent companies don't pay dividends at all. Many mediocre companies do, until they can't.

This is why a dividend should never be treated as an *entitlement*. It's a distribution of surplus cash. If surplus disappears, the dividend becomes negotiable. Sometimes it's reduced as a temporary measure; sometimes it's eliminated; sometimes it's "maintained" by borrowing money, which can look generous right up until it looks reckless.

Stocks can pay you in two currencies. The obvious currency is dividends. The less obvious currency is *option value*: a liquid stock gives you the ability to sell your future upside (covered calls) or get paid to set a buy price (cash-secured puts). Those options strategies are powered by collateral, and stocks are the most common form of collateral in retail accounts because they're standardized, liquid, and priced continuously.

That said, collateral isn't just "something the broker accepts." Collateral is what absorbs volatility so you can keep running the income machine without being forced to sell at the wrong time. A stock that drops 40% because the business is deteriorating is a very different kind of collateral than a stock that drops 20% because the whole market is having a bad quarter.

The investor's question isn't "what's the yield?" A high dividend yield can mean "generous." It can also mean "the price fell because the market smells trouble." The question that matters is: *what is the business earning, in cash, after keeping itself healthy—and is the dividend small enough to survive a rough year?*

You don't need to be a forensic accountant to think like an owner. You just need the right mental model: a dividend is a distribution from an engine, and engines need fuel and maintenance. If you strip too much cash out of the engine, it eventually breaks.

Bonds: lending with a contract (and a new kind of risk)

A bond is a loan you can trade. You hand over principal today. In return, the issuer promises periodic interest and repayment at maturity. That legal promise is comforting, but it introduces a different set of "gotchas" that many income investors learn only after they've built a bond-heavy portfolio and interest rates rise.

Bond income feels stable because the cash flows are scheduled. If you buy a traditional bond and hold it to maturity, the interest payments are usually fixed (unless it's a floating-rate or inflation-linked bond). You can plan around them. That predictability is why bonds sit at the emotional center of "income investing," especially for retirees who want a paycheck schedule.

But bond *prices* move every day, and they move for a reason that surprises equity investors: they move mainly because the discount rate changes.

Interest-rate risk is not a footnote—it's the steering wheel. Bond prices are tied to prevailing interest rates. When rates rise, existing bonds with lower coupons become less attractive, so their prices fall until their yield matches the new market reality. When rates fall, existing higher-coupon bonds become more valuable, so prices rise.

The longer the time until maturity, the more sensitive a bond tends to be to rate changes. This sensitivity is often summarized by *duration*, which is a rough measure of how much a bond's price could change for a 1% change in interest rates. You don't have to calculate duration by hand to use it; you just need to respect its message: long-term bonds can swing in price far more than "income investors" expect.

Credit risk: the contract is only as good as the borrower. Treasuries are backed by the U.S. government; investment-grade corporate bonds are backed by large companies; high-yield ("junk") bonds are backed by borrowers who pay more because lenders demand more compensation for risk. That compensation shows up as a higher yield, which is why high-yield bonds look so tempting in a yield screen.

But higher yield in credit markets often means one of three things:

- The borrower is more likely to default.
- The borrower is more cyclical (fine in booms, fragile in recessions).
- The bond has structural features that disadvantage you (callability, weak covenants, poor liquidity).

Call risk: the borrower can fire you as a lender. Many corporate bonds (and many preferred securities) can be "called," meaning the issuer can repay the bond early at a set price. If rates fall, the issuer has an incentive to refinance at a lower rate—great for them, annoying for you. You did your job: you lent money. You took the early risk. Then right when your bond becomes valuable, it gets redeemed and you have to reinvest at lower yields.

That's a quiet truth of lending: the borrower often has more flexibility than you do.

Liquidity risk: bonds don't always have a clean price tag. Stocks trade on centralized exchanges with tight spreads for popular names. Many bonds trade over-the-counter, with spreads that can widen dramatically when markets are stressed. This matters for collateral-minded income investing because "I can sell if I need to" is only true if the market is functioning and the instrument is liquid.

A bond that you can't sell without taking a haircut is not a great shock absorber, even if the coupon looks comforting.

Funds: the pre-packaged toolkits (ETFs and mutual funds)

If individual stocks are single machines and individual bonds are single loans, a fund is a factory floor in a box. Funds exist for one reason: most people want broad exposure without building it bolt by bolt.

There are two common wrappers:

- *Mutual funds:* generally priced once per day at net asset value (NAV), bought and sold through the fund company.
- *ETFs (exchange-traded funds):* trade intraday like stocks with bid/ask quotes, and can be bought or sold anytime the market is open.

For income investing, the key is that funds *pass through* the income generated by what they own—dividends from stocks, interest from bonds, sometimes capital gains distributions depending on the structure and turnover.

Funds solve concentration and complexity—but introduce "wrapper risk." The convenience is obvious: instant diversification, professional index construction (or management), automatic rebalancing, easy access to hard-to-buy markets, and usually low minimums.

The less obvious part is that you now have to understand the wrapper:

- What does the fund *actually* hold (and how does it change over time)?
- How much does it cost to run (expense ratio)?
- How does it trade (liquidity and bid/ask spreads for ETFs)?
- How does it distribute income (monthly, quarterly, irregular)?
- Does it use leverage or derivatives?

A fund can be a beautiful tool or a Trojan horse. Two "high-yield bond ETFs" can have wildly different credit quality. Two "dividend ETFs" can differ in sector exposure enough that one behaves like a defensive portfolio and the other behaves like a disguised bet on banks and energy.

ETF income is real—but execution costs are real too. Because ETFs trade like stocks, they come with a bid price (what buyers are offering) and an ask price (what sellers want). The difference—the bid/ask spread—is a cost you pay when you cross from one side to the other. For highly liquid ETFs the spread may be tiny; for niche or stressed products it can be meaningfully larger.

That spread matters to income investors because it's a cost that doesn't show up in the expense ratio and doesn't feel like a fee. It feels like "the market price." But it's a toll you pay at the moment you enter or exit.

Premiums and discounts: sometimes you pay more than the box is worth. An ETF has an underlying value based on the securities it holds—NAV, a per-share estimate of the basket's value. The market price can trade slightly above (premium) or below (discount) that value. In calm markets these gaps are usually small for big, liquid funds. In volatile markets—or in funds holding less liquid bonds—the gap can widen.

This is another reason to think like a factory owner. If you're buying a basket of assets as collateral for income strategies, you want to know whether you're paying a fair price for the basket today, not just what the basket holds in theory.

Collateral: the unglamorous ingredient that makes the income machine work

People talk about income strategies as if the income appears out of thin air—like clipping coupons from the sky. In practice, income strategies are built on collateral. Collateral is what you bring to the table that makes the transaction credible.

Dividends and interest are "natural yield." Stocks and bonds can generate cash without you selling shares. That's natural yield: dividends from equity ownership, interest from lending. It's the closest thing markets have to a paycheck that doesn't require liquidation.

But natural yield is only part of the income story because you can also get paid for *making commitments*—and commitments require collateral.

Options turn collateral into a working asset. When you sell a covered call, you're promising to sell your shares at a certain price if assigned. Your shares are the collateral. When you sell a cash-secured put, you're promising to buy shares at a certain price if assigned. Your cash (or T-bills, depending on the broker) is the collateral.

This is why the "toolbox" discussion matters. A portfolio built from volatile, fragile collateral forces you into defensive maneuvers at the worst times. A portfolio built from resilient collateral gives you choices: you can roll, wait, collect, and manage without panic.

Not all collateral is created equal. A few practical ways collateral can betray you:

- *Fragile business model:* a stock that looked safe until earnings collapse, and now the dividend and the price fall together.
- *Hidden leverage:* a fund that boosts yield with leverage, then de-levers in a drawdown, locking in losses.
- *Rate sensitivity:* a long-duration bond fund that drops sharply when rates rise, even though the "income" line item looks steady.
- *Liquidity mismatch:* a product that trades easily in normal times but becomes expensive to exit when everyone runs for the same door.

Collateral is supposed to be the sturdy floor under the machines. If the floor is made of trapdoors, your income plan becomes an anxiety plan.

Choosing the right tool for the job (without chasing shiny yields)

A useful way to think about these instruments is to match the tool to the type of income you're trying to build.

If you want income that can grow, equity is usually the growth engine. Dividends can rise over time if the business grows its cash flows. That makes dividend-paying stocks and equity income funds attractive for long horizons. The trade-off is that dividends are discretionary and stock prices can be volatile. You're paid to tolerate uncertainty.

If you want scheduled payments and defined maturity, bonds do that better than stocks. Bonds shine when you need cash flows you can map to a timeline (tuition, a down payment, bridging retirement). But you must respect duration and credit risk. Bonds are "safe" only in the specific sense that a high-quality borrower is likely to pay you back if you hold to maturity. In the meantime, market prices can move sharply.

If you want diversification and simplicity, funds are the fast lane. ETFs and mutual funds can give you broad exposure with one trade. They're often the easiest way to build a base layer of collateral. But don't let the wrapper make you lazy. Read what it holds, understand what drives its yield, and pay attention to trading costs and liquidity.

A quick reality check: income is not the same as safety

Income investing attracts people who want calm. The marketing language is soothing: "high income," "steady distributions," "enhanced yield." The numbers look comforting: 7%, 9%, 12%. But the market doesn't hand out high yield without collecting payment somewhere else—usually through volatility, credit risk, leverage, or the chance of a cut.

A dividend yield is a ratio, not a promise. A bond yield is compensation, not a gift. A fund distribution is a *policy outcome*, not a law of nature.

If you remember only one thing about the toolbox, remember this: the tool that looks the most generous on day one is often the tool that demands the most discipline on day 400.

Actionable takeaway: build your ``collateral core" before you optimize yield

Pick one holding you currently own (or plan to buy) for income—an individual stock, a bond, or a fund—and run this five-minute collateral check:

- Write down *why it pays you* (dividend policy, bond coupon, fund distributions).
- Write down *what would make the payment shrink* (earnings decline, default risk, rate shock, leverage unwind, distribution policy changes).
- Write down *what would make the price drop fast* (business fragility, duration, credit sensitivity, liquidity).

- Decide whether you'd be comfortable using it as *collateral* for a commitment (like selling a covered call or cash-secured put), even during a nasty quarter.
- If you can't explain the risk in plain English, size it smaller until you can.

That's how you keep the factory running: not by hunting the loudest yield, but by choosing tools that keep their shape under pressure.

2.2 Evaluating Business Quality

The market has a dark sense of humor: it often advertises the riskiest income with the biggest, brightest number.

A stock yielding 11% looks like a bargain until you realize the yield got there the same way a "half-off" sign gets there on a carton of milk that expires tomorrow—the price fell first. If the business is wobbling, a high dividend yield isn't a reward; it's a warning label with glitter on it.

Business quality is what separates a steady paycheck from a "surprise, we're cutting the dividend" press release. The good news is that you don't need to be an analyst with three monitors to spot the difference. The clues are already in plain sight, in documents that look intimidating mostly because they're written by committees and lawyers.

Think of it like buying a used car. The seller can talk all day about how smooth it drives. You want the maintenance records, the mileage, and a look under the hood. For a dividend payer, the "maintenance records" are the financial statements, the footnotes, and management's own explanation of what went right, what went wrong, and what could go wrong next.

The yield trap: a broken machine wearing a fresh coat of paint

A yield trap is an asset that looks like it's built for income but is quietly eroding the very thing that funds that income. The common patterns are surprisingly human:

- A mature company feels pressure to "keep shareholders happy," so it pays too much out even as the business weakens.
- Management borrows to fund dividends or buybacks because cutting the dividend feels like admitting defeat.
- A cyclical business gets valued as if the good times are permanent, then the cycle turns and the payout suddenly becomes impossible.

The trap isn't just the dividend cut. The trap is what usually travels with it: a sharp price drop, a tighter credit environment, layoffs, asset sales, and a management team suddenly talking about "strategic alternatives." Income investors don't merely lose the income stream; they lose the collateral that was supposed to support the strategy.

So the goal isn't "find the highest yield." The goal is "find the payment that can survive reality."

Where the truth hides: the 10-K and the parts most people skip

For U.S. public companies, the annual 10-K is the closest thing to a full mechanical inspection. You'll hear people say it's "too long," which is true in the same way it's "too long" to read a home inspection report before buying a house. The length is the point: it's comprehensive, standardized, and full of disclosures companies would not voluntarily put in a glossy investor deck.

Three parts do most of the heavy lifting:

- *Item 1: Business.* What the company actually does, how it makes money, and what it depends on.
- *Item 1A: Risk Factors.* A catalog of what can hurt the business or the stock-often more candid than earnings calls.
- *Item 7: MD&A (Management's Discussion and Analysis).* Management explaining results, liquidity, capital needs, and trends (sometimes with a little spin, but still revealing).

There's also a quiet fourth area that separates casual readers from people who sleep well at night:

- *Footnotes.* The place where the "simple" numbers get explained, qualified, and occasionally undermined.

If you're allergic to 10-Ks, don't try to read them cover-to-cover like a novel. Read them like a detective. You're not hunting for trivia. You're hunting for pressure points: dependence on one customer, refinancing needs, pension obligations, lawsuits, a business model that only works when money is cheap, or a "non-recurring" charge that seems to recur every year like clockwork.

A dividend can survive bad headlines. It has a much harder time surviving a balance sheet that's quietly running out of oxygen.

Metric #1: Free Cash Flow - the fuel in the tank

Earnings are an opinion; cash is a fact. That's a bit of Wall Street snark, but it points to something practical: dividends are paid in cash, not in accounting profits.

Free cash flow (*FCF*) is a simple concept with a lot of power. It's the cash a business generates after paying its operating bills and making the investments needed to keep the business running (often called capital expenditures, or "capex"). In plain English: after the company keeps the lights on and replaces the worn-out parts, how much cash is left over?

A business that consistently produces free cash flow has options: pay dividends, buy back stock, reduce debt, invest for growth, or build a cash cushion for ugly years. A business that doesn't produce free cash flow is often forced into uncomfortable choices: borrow, sell assets, issue stock, or cut distributions.

How to use FCF without drowning in math. You're not trying to build a spreadsheet empire. You're trying to answer a few high-impact questions:

- Is free cash flow *consistently positive* over a full cycle, not just in one good year?

- Does it look *stable*, or does it swing wildly with commodity prices, housing demand, ad budgets, or consumer credit?
- Does the company's own story about "investment" match reality, or is capex being deferred like skipping oil changes?

One of the most dangerous dividend stories is the company that "covers the dividend" only by under-investing. That can work for a while, and then it fails suddenly: equipment breaks, customer satisfaction drops, competitors leapfrog, and the business discovers that "maintenance capex" was not optional after all.

A quick smell test in MD&A. When management talks about cash flow, listen for what they *blame*. If cash flow fell because they built inventory ahead of demand, that might reverse. If cash flow fell because customers are paying slower, that can be a credit problem. If cash flow fell because margins compressed and stayed compressed, that's often a competitive problem-which tends to be stickier than a single bad quarter.

A high-quality income business doesn't need heroic explanations for why cash keeps showing up.

Metric #2: Payout ratios - the pressure gauge on the dividend

A payout ratio answers a blunt question: how much of the company's earnings (or cash flow) is being handed out to shareholders?

There are multiple ways to calculate it. The two most useful for income investors are:

- *Earnings payout ratio:* dividends divided by earnings per share.
- *Cash payout ratio:* dividends divided by free cash flow (or "free cash flow to equity," depending on the reporting).

The earnings version is easy to find, but it can be distorted by one-time charges, non-cash items, or accounting quirks. The cash version is often more honest, because the dividend leaves the building as cash.

Here's the key behavioral insight: a payout ratio isn't just a number; it's a statement about management's posture.

A moderate payout ratio says: "We can pay you and still reinvest, absorb shocks, and keep our balance sheet healthy." A very high payout ratio says: "We're living close to the edge, and we're betting nothing goes wrong."

That edge can be fine in certain industries-some utilities and REIT-like structures are designed to pay out a large share. But for ordinary operating businesses, persistently high payout ratios usually mean one of two things: limited reinvestment opportunities (which can be okay) or financial strain (which is not).

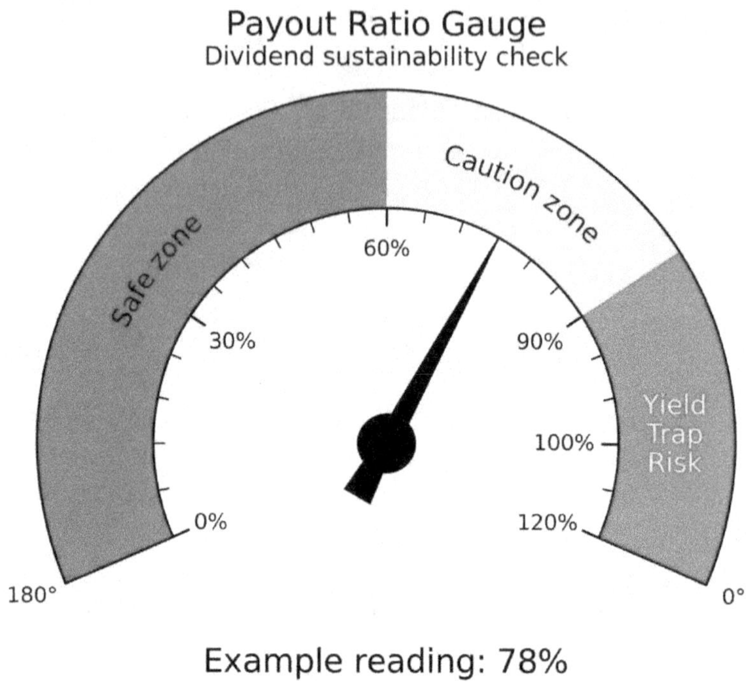

What payout ratios feel like in real life. A payout ratio in the "safe" range gives you room for error. A bad quarter doesn't force a decision. A recession might sting, but it doesn't automatically break the promise.

In the caution range, you start needing everything to go "mostly right." The business can still be fine, but the dividend now competes with debt reduction, reinvestment, and basic resilience.

In the danger range, the dividend is often a hostage to sentiment. Management might keep paying to avoid panic, even if it weakens the company. Or management might cut abruptly because it has no choice. Either way, an income investor's worst day often arrives when a high payout ratio meets a real-world shock: higher rates, lower demand, a key customer leaving, input costs rising, or a credit downgrade that makes refinancing expensive.

A practical rule: watch the direction, not just the level. A payout ratio drifting upward over several years is often more ominous than a one-year spike. It can signal margins compressing, capex rising, or management refusing to adjust the dividend to match reality. Healthy companies usually keep the payout ratio within a deliberate band. Unhealthy ones let it creep, then defend it with increasingly creative language.

Metric #3: Economic moat - the reason cash flow doesn't evaporate

An economic moat is a simple idea: it's what keeps competitors from copying the business and stealing its profits. If free cash flow is the fuel and the payout ratio is the pressure gauge, the moat is the fence around the fuel depot.

Without a moat, a dividend can be a temporary artifact of good times. With a moat, a dividend can become a habit-something the company can sustain because customers stick around and pricing power holds up.

Moats come in flavors you can recognize without jargon:

Switching costs. Some products are painful to replace. Businesses pay for software that is deeply embedded in workflows. Hospitals standardize on certain systems. A manufacturer qualifies a supplier and doesn't want to repeat the process. Switching costs don't have to be dramatic to be powerful; they just have to be annoying enough that customers stay.

Network effects. A platform becomes more valuable as more people use it. Think payment networks, marketplaces, or communication systems. Network effects are rare and powerful, and when they exist, they can support durable cash flows-though they can also attract regulatory scrutiny.

Cost advantage. Some businesses can produce or deliver more cheaply than competitors, due to scale, logistics, or process discipline. A cost advantage is a moat that tends to shine during downturns: when everyone's margins get squeezed, the low-cost operator survives and often gains share.

Brand and trust. Brand sounds fluffy until you watch consumers pay more for the same category of product because they trust the label. Trust is especially valuable in areas where failure is costly: finance, healthcare, safety-related products, and mission-critical services.

Regulation and licensing. In some industries, the moat is a license, a permit, a regulated monopoly, or a long approval timeline. These moats can be strong, but they also come with political risk. A dividend built on regulatory goodwill needs a margin of safety because the rules can change.

A quality checklist hidden in plain sight

You can often infer moat strength and business fragility from a few recurring lines in filings and calls-not because management is confessing, but because they can't avoid describing reality.

Concentration risk. If a company depends on one customer, one supplier, one geography, or one product line, the dividend is effectively tied to a single point of failure. In the 10-K, look for disclosure of major customers, and watch for phrases like "a significant portion of revenue."

Pricing power language. Businesses with pricing power talk differently. They describe "pricing actions" that stick, stable gross margins, and demand that holds up. Businesses without pricing power talk about "competitive pricing pressure," "promotional intensity," and "mix shift" as if it's weather.

Debt maturity wall. A dividend can look safe right up until the company has to refinance a pile of debt at much higher interest rates. In the footnotes and liquidity discussion, watch the maturity schedule. If large amounts come due in the next few years, the dividend is competing with survival.

Share count discipline. A company that constantly issues stock to fund dividends (directly or indirectly) may be paying you with one hand while taking your ownership with the other. Dilution isn't always bad-sometimes it funds growth-but persistent dilution paired with a high dividend deserves suspicion.

"Adjusted" everything. Non-GAAP measures can be useful, but the pattern matters. If "one-time" charges happen every year, they are not one-time; they are the business. The dividend doesn't care how earnings are labeled-it cares whether cash is there.

How quality shows up in dividend behavior

Dividend history isn't a guarantee, but it's a personality test.

Healthy dividend payers treat the dividend like a long-term promise. They don't set it at the maximum possible level. They leave room. They increase gradually. They prioritize resilience, because they know the market punishes cuts disproportionately.

Fragile dividend payers treat the dividend like a marketing tool. They aim for a headline yield. They defend it even as fundamentals soften. They rely on favorable conditions-low rates, strong commodity prices, benign credit markets-to keep the math working. When conditions change, the dividend becomes the first casualty or the last stubborn stand before a cut.

One of the most revealing moments is when a company faces a difficult year. Quality management teams talk about protecting the balance sheet, investing through the cycle, and preserving flexibility. Low-quality teams talk about "commitment to the dividend" with the intensity of someone trying to talk themselves into it.

A small, powerful habit: connect the dividend to the business story

A dividend is never just a number. It's an output of a business model.

So each time you consider an income asset, force a simple sentence:

> "This company can keep paying because ______________________________."

If the blank is filled with something fragile-"because oil stays above $X," "because consumers keep borrowing," "because refinancing stays easy," "because no new competitors show up"-you're not analyzing a dividend. You're betting on a condition.

If the blank is filled with something durable-"because switching costs are high," "because costs are structurally lower," "because demand is regulated and predictable," "because the balance sheet is conservative"-you're closer to a dividend that can survive the messy parts of capitalism.

Actionable takeaway: a 20-minute dividend durability test

The next time a high yield catches your eye, run this quick test before you fall in love with the number:

1. Pull up five years of cash flow and answer: *Was free cash flow positive in most years, and does it look repeatable?*
2. Check dividend coverage: *Is the dividend comfortably covered by free cash flow in a normal year, not just by "adjusted" earnings?*
3. Read the 10-K with a highlighter mindset:
    - Item 1A (Risk Factors): circle the risks that would directly hit cash flow or refinancing.
    - Item 7 (MD&A): find the paragraph on liquidity and capital resources; note any dependence on capital markets.

4. Skim the footnotes for "quiet obligations": debt maturities, leases, pension/retirement commitments, and any big contingent liabilities.
5. Write one sentence on moat: *Why can this business raise prices or keep customers without bleeding market share?* If you can't write it, size the position as if the moat is weak-because it might be.

If those five steps feel like work, that's the point. Yield is a product label. Business quality is the ingredient list. Only one of them tells you what you're actually consuming.

2.3 Market Mechanics and Execution

Buying a stock feels like buying a toaster—type the ticker, hit the button, and you own it. The screen even gives you "last price," like there's a sticker on the shelf.

That's not what's happening.

A modern market is a nonstop auction where prices flicker because thousands of participants disagree, in real time, about what something is worth *right now*. Your broker is not a store. It's a doorway into that auction. And every time you step through, you pay a toll that isn't listed as a fee: the mechanics of execution.

Income investing makes those mechanics matter more, not less. If you're collecting dividends and selling options, you're doing many small transactions over time. A little "hidden tax" on each trade quietly becomes a meaningful drag on your factory's output—especially when you trade less liquid names, use the wrong order type, or hit the button during a volatility spike.

The bid and the ask: the two price tags that actually exist

When you pull up a quote, you'll see two prices:

- *Bid:* the highest price someone is currently willing to pay.
- *Ask (or offer):* the lowest price someone is currently willing to sell for.

The *bid-ask spread* is the gap between them. If the bid is $50.00 and the ask is $50.05, the spread is $0.05.

That gap is not a rounding error. It's the market's way of saying: "Sure, you can trade instantly—but immediacy isn't free."

If you buy immediately, you typically buy at the ask. If you sell immediately, you typically sell at the bid. Cross the spread and you've paid the toll.

Why spreads exist (and why you should care). Spreads compensate the middle of the market for providing liquidity—standing ready to buy from sellers and sell to buyers. Sometimes that "middle" is a market maker. Sometimes it's a high-frequency trading firm. Sometimes it's another investor who happens to have a resting order.

Whoever it is, they are taking a risk: the price can move against them between the time they commit and the time they unwind. The spread is how they get paid for taking that risk.

Your job is not to hate spreads. Your job is to recognize when they're tiny (a small toll) and when they're huge (a highway robbery with a polite interface).

The hidden tax you pay without noticing

Suppose you buy 100 shares and the spread is $0.05. If you cross it, that's about $5 of "cost" right away. Not tragic.

But now imagine you run an income approach that trades repeatedly: buying, trimming, rolling options, rotating tickers, rebalancing. Or imagine you're trading options where spreads can be wider than the underlying stock. Or imagine you're trading a thin ETF where the spread isn't $0.05—it's $0.30, $0.60, sometimes more.

Now that toll starts to resemble a subscription fee you never signed up for.

The painful part is psychological: you don't get an email that says "Spread charged: $37.50." You just start your position slightly underwater. It feels like bad luck. It's actually microstructure.

A quick way to "see" the spread cost. If a stock quote shows:

$$\text{Bid} = 50.00 \quad \text{Ask} = 50.10$$

and you hit a market buy, you're likely paying about $50.10. If the "true" midpoint is $50.05, you've effectively paid $0.05 per share above the midpoint. That's the hidden tax of demanding immediacy.

The midpoint isn't a magical fair price, but it's a useful reference point: if you can consistently execute near the midpoint, you've built yourself a quiet edge.

Liquidity: the difference between a deep pool and a puddle

Liquidity is a fancy word for a simple thing: how easily can you trade *size* without moving the price?

A highly liquid market is like a deep pool. You can cannonball in and the water barely moves. A low-liquidity market is a puddle. You step in and the whole thing splashes.

Liquidity shows up in a few ways:

- *Tighter spreads:* more competition to provide liquidity compresses the toll.
- *More depth:* more shares available at multiple price levels.
- *More stable execution:* your order is less likely to "walk the book" and fill at progressively worse prices.

Income investors should have a bias toward deep pools. You can still fish in the puddles, but you need to do it carefully—with smaller orders, limit prices, and extra patience.

Volume is not the whole story. People use volume as shorthand for liquidity, and it helps, but it's not perfect. A stock can trade a lot and still have weird moments. An ETF can trade actively but hold less liquid assets underneath (certain bond ETFs are the classic example). In calm markets it feels liquid. In stressed markets the spread can widen and the price can drift away from the value of the underlying holdings.

That's not a moral failing of ETFs; it's a reminder that the wrapper trades at the speed of the stock market even when the underlying assets trade at the speed of paperwork.

Market orders: writing a blank check

A *market order* means: "Fill me now at the best available price."

The convenience is seductive. The risk is subtle: you are surrendering price control. In a fast market, the "best available price" can change between the moment you click and the moment you fill. If liquidity is thin, your order can fill across multiple price levels—especially if you trade more shares than the quote can absorb.

That's why a market order is essentially a blank check. It's not always a dangerous tool, but it becomes dangerous exactly when you most want to trade—during volatility.

The bad fills that happen when you're not looking. You place a market buy after a headline hits. The quote you saw shows an ask of $25.20. But the ask evaporates because someone else bought it first, or because liquidity providers widen spreads to protect themselves. Your order chases the price and fills at $25.60. You're instantly down 1.6% and you have no clue why, because the chart shows a blur and your brain labels it "slippage."

Slippage isn't random. It's the price of urgency.

If your strategy is built on lots of small edges—dividend yield plus option premium—then routinely overpaying by 0.5% or

1% on entries is like punching holes in your own cash-flow bucket.

Limit orders: choosing your price, not your luck

A *limit order* means: "Fill me at this price or better."

Now you have price control. The tradeoff is emotional: you might not get filled. The market might move away from you, leaving you with no position and a nagging feeling that you "missed it."

That feeling is exactly why limit orders work. The market constantly tempts you to pay up for immediacy. A limit order is you saying, politely, "No thanks. I'll buy if the price comes to me."

How to place a limit order like a pro (without pretending you're a pro). A practical habit is to use the midpoint as a starting point. If the bid is $50.00 and the ask is $50.10, consider a buy limit around $50.05. You're offering liquidity rather than demanding it.

Sometimes you'll get filled quickly. Sometimes you won't. If you don't, you can adjust—slowly—rather than chasing.

The same logic applies to selling. If you want out, a sell limit near the midpoint can reduce the toll you pay.

When a limit order can be worse than a market order. If you set an unrealistic limit because you're anchored to a price, you can end up never trading while the opportunity passes. The right use of a limit order isn't stubbornness; it's discipline.

You're deciding what edge is acceptable, not refusing to engage with reality.

Stop orders: the trapdoor that opens into a market order

A *stop order* is often described as protection: "If it drops to this price, sell." The catch is that many stop orders become *market orders* once triggered.

That means the moment you most want control—when prices are moving fast—you may lose it.

Stops can still be useful tools, but you should treat them like emergency exits that sometimes lead to a staircase under construction. In a sharp downdraft, the stop triggers, becomes a market order, and fills at a price much worse than your stop level. The phrase for this is "gapping through your stop." It feels unfair. It's just mechanics.

Income strategies that rely on holding collateral and managing positions tend to prefer planned exits and position sizing over automatic panic buttons. If you use stops, use them with full awareness of what they really are in a fast market.

The bid-ask spread in pictures: who pays, who waits, who bridges

The market can feel abstract until you visualize the two crowds—buyers and sellers—standing apart, and the market maker bridging the gap. The key execution difference is simple: a market order runs across the gap; a limit order waits at a chosen price level.

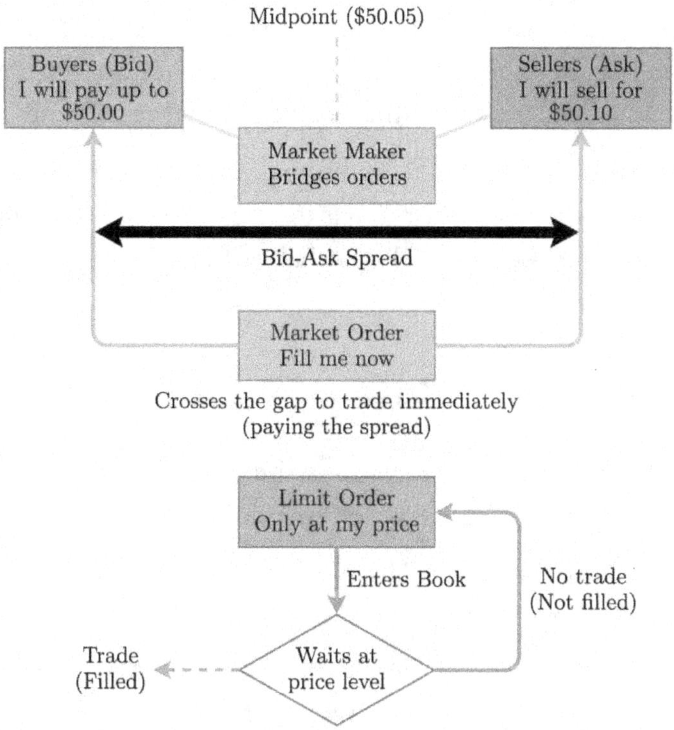

Order routing and ``best execution": your broker is making choices

When you place an order, it doesn't simply float into the market like a message in a bottle. It's routed—sent to a market center or liquidity provider. Brokers have a duty called *best execution*, meaning they're expected to seek the most favorable terms reasonably available, considering factors like price, speed, and likelihood of execution.

That's reassuring, but it's not magic. "Best" involves tradeoffs. A routing choice that maximizes speed might not maximize price improvement. A choice that improves price for small orders might behave differently for larger ones. And arrangements like payment for order flow can complicate incentives even when the broker is still meeting formal requirements.

For a practical income investor, the point isn't to become a microstructure lawyer. The point is to act like execution matters—because it does—and to choose habits that make good execution more likely.

ETFs: the spread you see and the spread you don't

ETFs trade like stocks, so you see the bid and ask. But there's an extra layer: the ETF's underlying holdings.

In very liquid equity ETFs, the underlying stocks are also liquid, so the ETF price tends to track its basket tightly and spreads stay narrow. In ETFs holding less liquid bonds, bank loans, or niche assets, the underlying market may be slower and fuzzier. In calm markets, everything looks fine. In stressed markets, spreads can widen and the ETF can trade at a premium or discount to its net asset value.

If you use ETFs as your income base or collateral layer, trade them like you respect the wrapper. Wide spreads and volatile premiums/discounts are not just trivia; they're execution costs.

Practical execution habits that protect your income edge

A good trade isn't just a correct idea. It's a correctly *executed* idea. Execution is where people donate money to the market without realizing they've made a donation.

Trade when the market is awake. Spreads are often wider at the open and near the close, and during fast news. Many liquid names still trade fine, but if you're trading something thinner, patience can be profitable.

Use limit orders by default. If you're not in a true emergency, price control is your friend. Even a simple midpoint limit can materially reduce the toll you pay over hundreds of trades.

Break up size if liquidity is questionable. If your order is large relative to the typical size you see on the quote, consider slicing it. You're trying to avoid stomping into the puddle and splashing the price against yourself.

Watch spreads on the option, not just the stock. Options can have much wider spreads than their underlying stocks, especially far from the money or in less popular expirations. If you're selling premium, a wide spread is the market charging you rent for the privilege.

Don't confuse "commission-free" with "cost-free." Most people learned to fear commissions. The market learned to hide costs elsewhere—spreads, slippage, and execution quality. The invoice is still coming; it just arrives as a worse price.

Actionable takeaway: a simple execution checklist before you click

Before your next buy, sell, or options trade, pause for ten seconds and answer four questions:

1. What's the spread right now (in dollars and as a percentage)?
2. Is this instrument liquid enough that a market order is safe, or am I about to write a blank check?
3. Can I place a limit near the midpoint and let the market come to me?
4. If this is an ETF or option, am I looking at a wrapper that could have wider spreads or weird pricing during stress?

If you do nothing else but switch from "click and pray" to "limit and negotiate," you'll keep more of your edge. The market will still be an auction, but you won't be the person walking in and shouting, "I'll pay whatever."

2.4 Brokerage Accounts and Settlement

The first time an investor gets a "good faith violation" warning in a cash account, it feels like being pulled over for speeding on a road with no speed-limit signs. You sold something, you bought something else, and suddenly your broker is scolding you—despite the fact that the money is obviously "there," at least on your screen.

That mild confusion is your introduction to the plumbing.

Markets are not just prices and tickers; they're a set of promises, timestamps, custody rules, and tax containers. If you're building an income approach—especially one that will eventually involve frequent small trades, option premium, and dividend cash flows—then the account you choose and the way trades settle can either feel like a smooth conveyor belt or like trying to run a factory where the doors lock at random.

None of this is glamorous. All of it matters.

The broker is your warehouse, not your advisor

A brokerage account is a warehouse for financial assets. It holds your shares, your bonds, your fund positions, your cash, and (if you use them) your option contracts. It also keeps score: cost basis, realized gains, tax forms, and the daily mark-to-market value that can make you feel richer or poorer before breakfast.

A subtle but important mental shift: your broker is not merely a "platform." It is a custodian and a credit gatekeeper. It decides what counts as buying power, what counts as settled cash, what you can pledge as collateral, and what you are allowed to do when a trade hasn't fully finished its journey through the clearing system.

You can run an income strategy with an average broker, but you can absolutely sabotage one by misunderstanding how the warehouse doors work.

Trade date versus settlement date: the handshake and the delivery

When you click *Buy* or *Sell*, the trade is executed immediately in the market. That's the handshake—trade date, often abbreviated T.

Settlement is the delivery. It's when the cash and the securities officially change hands in the back office—settlement date. In the U.S., standard settlement for most stocks and ETFs is $T+1$, meaning one business day after the trade.

That "+1" sounds tiny until you try to reuse money quickly.

A concrete timeline that actually happens. If you sell a stock on a Monday, the trade typically settles on Tuesday (assuming normal market hours and no holidays). The position disappears from your holdings right away, but the cash you receive is not always treated as fully settled and reusable until settlement.

This is why the market can feel like it's both instantaneous and delayed at the same time. The auction clears instantly; the paperwork clears on schedule.

Why settlement exists at all. Settlement is the system preventing chaos. It's how the market ensures that when you sell, you actually deliver shares, and when you buy, you actually deliver cash. Shortening settlement to $T + 1$ (effective May 28, 2024 in the U.S.) was aimed at reducing credit and market risk in the system—the less time between handshake and delivery, the less time something can go wrong in between.

For you, the practical message is simpler: your cash account has a rhythm, and the rhythm is one business day.

Cash accounts: ``I paid," but did I *settle*?

A cash account is exactly what it sounds like: you can only buy securities with cash you actually have. No borrowing. No leverage. It's the cleanest setup for many income investors because it forces discipline and reduces the chance of an accidental blow-up.

But a cash account has a quirk that surprises people: *unsettled funds*.

Unsettled funds are not imaginary money; they're in transit. When you sell a security in a cash account, the proceeds are typically not considered settled until the settlement date. If you use those proceeds to buy something else and then sell the new thing before the original sale settles, you may trigger restrictions (brokers often call these "good faith violations" or similar).

It's not about morality. It's about the broker ensuring you're not effectively free-riding—cycling trades without settled payment.

How this interacts with income trading. If you plan to do frequent adjustments—rolling options, trimming positions, re-entering—settlement timing shapes your cadence. $T+1$ helps, but it doesn't remove the concept. If you're running a strategy that expects rapid reuse of proceeds, you need to plan around settlement the way a small business plans around invoices getting paid.

A simple habit helps: treat unsettled cash as "pending," even if your app makes it look spendable.

Margin accounts: borrowing power, plus new ways to get hurt

A margin account lets you borrow against your securities. It can increase flexibility: faster reuse of proceeds, the ability to hold positions while deploying cash elsewhere, and easier handling of short option positions (many brokers require margin approval for certain option strategies).

Margin is not automatically reckless. But it is automatically *sharper*. It turns your warehouse into a warehouse with forklifts—more productivity, more risk of losing a toe.

The two margin risks people underestimate. First, *interest*. Borrowing costs money, and that cost can quietly eat the edge of an income strategy. If you're collecting a 6% yield but paying 8% margin interest on borrowed funds, you've built a machine that runs backward.

Second, *forced selling*. If your account value drops enough relative to what you've borrowed, the broker can require you to deposit more cash or liquidate positions. That's the classic margin call. It tends to happen when markets are already down—when you least want to sell.

Income investing often aims to reduce the chance of being forced to sell at the wrong time. Margin increases that chance unless used carefully and conservatively.

Regulation T: the rulebook behind the curtain. U.S. margin mechanics are shaped by Federal Reserve Regulation T, which governs how broker-dealers extend credit and the timing of payments. You don't need to memorize Reg T percentages to benefit from this fact; you just need to recognize that the broker's restrictions are not arbitrary. They're part of a broader credit-control system.

If a broker says "you can't do that," it's rarely personal. It's the warehouse following fire code.

``Street name" and custody: where your shares actually live

Many investors imagine their shares sitting in a digital drawer labeled with their name. In practice, most securities are held in *street name*. That means the broker (or its nominee) is listed as the registered owner on the issuer's books, while you are the beneficial owner with rights to the economic value.

This setup makes modern trading possible at scale. It also introduces a different category of risk—*custody risk*—which is not the same as market risk.

SIPC: protection against brokerage failure, not market loss. In the U.S., many brokers are members of SIPC (Securities Investor Protection Corporation). SIPC protection is designed to help customers recover missing cash and securities if a SIPC-member brokerage fails—up to $500,000 per customer, including up to $250,000 for cash. It does *not* protect you from your stock going down, your bond fund losing value, or your options trade being wrong.

This distinction matters because income investors sometimes seek "safety" in the wrong dimension. SIPC is about your warehouse burning down, not about your inventory losing value.

The practical behavior that follows. Use reputable, SIPC-member brokers. Keep your own records. Pay attention to cash balances. And don't confuse a broker's glossy interface with a guarantee. Your real protection is understanding what is and isn't protected—and sizing risks accordingly.

Three account types that change the after-tax reality

Two portfolios can generate the same pre-tax cash flow and deliver radically different results after taxes. The container matters.

Most income investors will interact with three common account types:

- *Taxable brokerage account*
- *Traditional IRA (tax-deferred)*
- *Roth IRA (tax-free qualified withdrawals)*

Each container has a personality. The mistake is treating them as interchangeable parking lots. They are more like different factory zones with different rules about what you can store, what you can move, and what the government takes when you ship the product out.

Taxable accounts: flexible, but every distribution has a shadow

A taxable brokerage account is the most flexible. You can generally withdraw money any time without early-withdrawal penalties. You can tax-loss harvest. You can mix strategies freely. For many investors, it's the main "operating account" of an income approach.

The tradeoff is simple: the IRS is your silent partner, and it wants updates all year.

Dividends: not all are taxed the same. Some dividends are "qualified" and may receive favorable tax rates; others are taxed as ordinary income. Funds can distribute a mix. The details depend on holding period and the nature of the payer. The result is that "the same yield" can land differently after tax.

Options premium: income in feel, complexity in taxation. Option premium feels like rent. Tax reporting can feel like paperwork. Short-term trading gains are generally taxed at ordinary rates. Some index options can have special tax treatment. Wash sale rules can complicate active trading. The main point isn't to become a tax expert overnight; it's to recognize that a high-frequency premium strategy in a taxable account may produce a lot of short-term taxable events.

If you're building a cash-flow machine, taxes are a leakage point. You don't ignore leaks.

Traditional IRA: a tax bunker with a locked door

A Traditional IRA is a tax-deferred container. Contributions may be tax-deductible depending on your situation, and investments grow without annual taxation inside the account. You generally pay taxes when you withdraw.

For income investors, the appeal is obvious: you can reinvest dividends and option premium without the annual tax bite, letting compounding work more efficiently.

The tradeoff is access. Withdrawals are governed by age rules, and there are penalties for early withdrawals in many cases. It's not an everyday operating account; it's more like a long-term vault.

Strategy fit: where tax deferral shines. Tax-deferred accounts can be excellent for higher-turnover income strategies because they reduce the drag from frequent taxable events. But you still face brokerage rules: option approval levels, margin permissions, and what strategies are allowed in that specific IRA. Many brokers restrict certain advanced strategies in retirement accounts for good reasons.

So the IRA can be a bunker, but it still has safety protocols.

Roth IRA: the cleanest compounding, with the strictest funding rules

A Roth IRA is the opposite flavor: you contribute after-tax dollars (subject to eligibility rules), and qualified withdrawals can be tax-free. That makes it uniquely powerful for compounding, especially when the strategy involves growth plus income.

A Roth is emotionally satisfying because it makes the math feel honest: what you see is closer to what you keep.

But it has constraints: contribution limits, eligibility, withdrawal rules, and again, broker-level restrictions on what strategies you can run. Many investors treat the Roth as the place for the highest expected long-term return per unit of risk—because shielding growth from future taxes can be tremendously valuable.

Putting strategies in the right container: a practical way to think about it

Rather than trying to memorize tax rules, use a simpler decision lens: *What kind of "income" is this, and how often does it generate taxable events?*

Low-turnover dividend investing in taxable can be fine. If you own quality dividend stocks or broad ETFs and you're not trading much, a taxable account can be perfectly reasonable. You keep flexibility, and the tax reporting is manageable.

Higher-turnover premium selling often benefits from shelter. If you plan to actively sell options, roll positions, and generate frequent short-term gains, sheltering that activity in an IRA (where allowed) can reduce annual tax friction. Whether the Traditional or Roth is better depends on your situation, but the general principle holds: frequent taxable events are heavier in taxable accounts.

Bonds and bond funds: watch the character of the income. A lot of bond interest is taxed as ordinary income in a taxable account. That can make after-tax yields less exciting than they look. Tax-sheltered accounts can make bond income more efficient for many investors, though specific circumstances vary.

The theme is not "avoid taxes at all costs." The theme is "don't accidentally choose the most leaky container for the most tax-inefficient income."

Settlement meets strategy: why timing changes how you manage positions

If you're building an income process that includes dividends, rebalancing, and options management, settlement timing turns into a very practical constraint.

Rolling options and moving collateral. When you roll an option (closing one and opening another), you may create multiple trades in quick succession. Depending on the account type and broker rules, you might need margin approval, option level approval, or sufficient buying power. If proceeds are unsettled or collateral is tied up, you can find yourself "right idea, wrong plumbing."

Dividend dates and cash planning. Dividends have ex-dividend dates, payment dates, and settlement interactions if you're buying or selling around those dates. You don't need to become a dividend-date sniper, but you do need to plan cash needs. If you rely on dividend cash to fund purchases, remember that dividends arrive on a schedule, not on your preferred day.

The emotional benefit of understanding the machinery. A lot of investor stress comes from interpreting normal plumbing as personal danger. Settlement delays, buying-power changes, and restriction warnings feel like disaster when you don't understand them. Once you do, they become logistics—annoying sometimes, but manageable.

And that calm is not a small advantage. Calm is what lets you keep making good decisions when prices are loud.

Actionable takeaway: set up your accounts like a real operation

Do this once, and you'll stop tripping over avoidable frictions:

1. *Label each account by purpose.* One sentence: "This account is for ____________________." (Long-term dividends, active options, bond income, etc.)
2. *Turn settlement into a habit.* In a cash account, assume sale proceeds are usable *next business day* ($T+1$) unless your broker shows them as settled.
3. *Decide where high-turnover activity lives.* If you expect frequent option rolls and short-term gains, consider whether a tax-sheltered account (where permitted) reduces leakage.
4. *Confirm your protections.* Verify your broker is a SIPC member, understand that SIPC is about custody failure (missing assets), not market losses, and keep a simple external record of holdings and cash.
5. *Keep a small cash buffer.* A little idle cash feels inefficient until it prevents you from selling the wrong thing on the wrong day because settlement timing boxed you in.

A great income strategy isn't only about picking good assets and executing trades well. It's also about making sure the warehouse doors open when you need them to—and that the tax bill doesn't quietly steal the shift's output.

Chapter 3

The Dividend Engine

Think of your portfolio as a collection of rental properties, but without the leaky faucets or 3 AM tenant calls. Dividends are your rent checks-predictable, cash payments that arrive simply because you own a piece of a business. But not all checks are created equal. In this chapter, we lift the hood on the dividend engine to understand how these payments work, how to spot a company that's writing checks its bank account can't cash, and how to tune your engine for either maximum immediate horsepower or long-term reliability.

3.1 Dividend Mechanics: The Calendar of Cash

Imagine you walk into a movie theater five minutes after the lights go down. The seats are still there. The screen is still there. The popcorn is still wildly overpriced. But the *one thing you came for*—the opening scene—already happened. Dividends work the same way. You can own the stock and still miss the payment if you show up on the wrong day.

That's why dividend investing isn't just about *what* you buy. It's about *when you own it*. There's a small, rigid calendar behind every dividend, and the market treats that calendar like airline boarding rules: you can argue with the gate agent if you want, but the plane is leaving anyway.

The good news is that once you understand the timeline, dividend mechanics stop feeling mysterious and start feeling mechanical—like a well-labeled machine. Four dates matter for a plain-vanilla cash dividend:

- *Declaration date*: the company announces the dividend.
- *Ex-dividend date*: the cutoff day for *who gets paid*.
- *Record date*: the company checks its shareholder list.
- *Payment date*: the cash actually lands.

Most confusion comes from one simple mistake: people assume the *payment date* is the important date. It feels like it should be. It's when the money shows up. But in dividend land, the payment date is just the mailman. The decision about *who* receives the cash is effectively made earlier, at the ex-dividend cutoff.

Declaration Date: The Company Makes a Promise (Out Loud)

The declaration date is the corporate equivalent of someone standing up at a dinner party and clinking a glass.

Management (through the board of directors) announces: "Here's the dividend, here's who qualifies, and here's when we'll pay it." This is where the dividend amount becomes real, public, and tradable information. Before declaration, the market may *expect* a dividend; after declaration, the market *prices* a dividend.

Two practical details hide inside that announcement:

- The company names a *record date* (who must be on the books).
- The exchange rules determine the *ex-dividend date* (the trading cutoff).

If you've ever watched a stock pop a little after a dividend announcement, you've seen the market reacting not just to the cash amount, but to what the announcement implies: stability, confidence, and the willingness to commit future cash flow.

And then, quietly, the clock starts ticking.

Ex-Dividend Date: The Cutoff That Actually Matters

If you only tattoo one dividend concept on your brain, make it this:

Buy before the ex-dividend date, and you get the dividend. Buy on the ex-dividend date (or later), and you don't.

That rule is the investor-facing reality, even though the company technically uses the record date to check its shareholder list. In U.S. markets, the ex-dividend date is set by exchange rules based around settlement timing, and it's typically set as the *record date* (or one business day before if the record date is not a business day). (investor.gov)

Why does this cutoff exist at all? Because trades aren't instantaneous at the plumbing level. Even though your brokerage app updates your position immediately, the official handoff of shares and cash happens on a settlement cycle. The U.S. market shifted to a $T + 1$ settlement standard (trade date plus one business day) with a compliance date of May 28, 2024.

(sec.gov) That shorter cycle is great for reducing risk in the system, but it also means dividend entitlement needs a clean, standardized line in the sand.

Here's the "show up late to the theater" translation:

- If you buy *before* the ex-dividend date, you arrived in time. You're treated as the buyer who gets the upcoming cash.
- If you buy *on* the ex-dividend date, you arrived after the curtain fell. You own the stock now, but the dividend belongs to the seller.

One more detail that surprises people: you can usually *sell on the ex-dividend date* and still receive the dividend. The cutoff is about whether you owned shares *before* the stock starts trading "ex-dividend" (literally, "without the dividend").

And this is where the "free money" myth dies.

On the ex-dividend date, the market mechanically re-labels the shares: *these shares no longer come with the next dividend attached*. All else equal, that makes the shares worth about one dividend less. So the price typically adjusts downward by roughly the dividend amount when the stock goes ex-dividend. That's not a moral statement. It's not punishment. It's just arithmetic: the company is about to send cash out the door, so the equity is about to be worth less cash. (investopedia.com)

If a stock is at $50 and it goes ex-dividend for $0.50, a "clean" textbook world would have it open around $49.50 the next morning. Real markets are messier—news happens, indexes move, buyers and sellers disagree—so you won't see a perfect one-for-one drop every time. But the direction and logic are the point: the dividend is not a magic coupon that appears out of thin air. It comes out of the company's value.

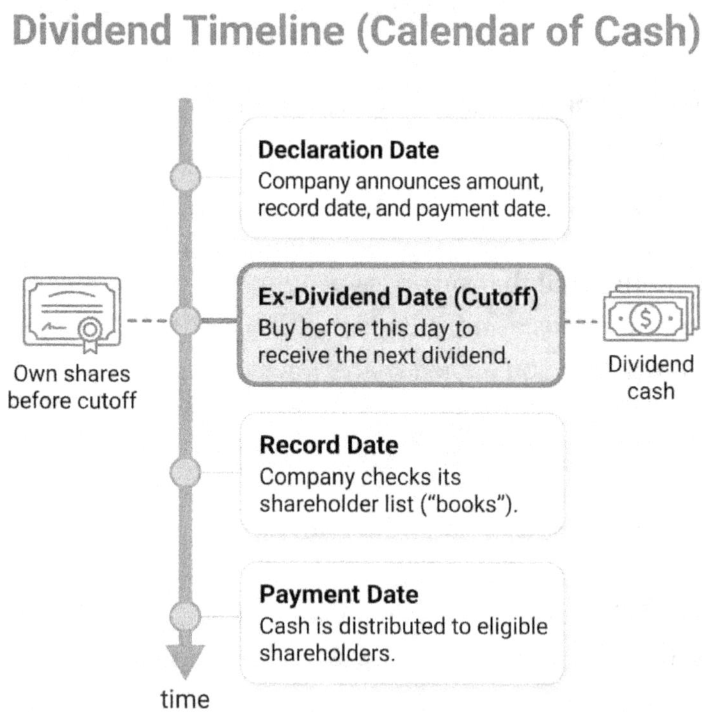

Record Date: The Roll Call (Mostly Administrative)

The record date is when the company (or, more precisely, its transfer agent and the clearing system) does a roll call: "Who counts as a shareholder for this dividend?"

If you're investing through a brokerage account, your name typically doesn't appear directly on the company's shareholder register. Shares are often held in "street name" through intermediaries. That sounds shady, but it's just how modern markets keep settlement and custody efficient at scale. Still, the logic remains: the company needs a defined moment to say who is entitled.

Here's the key: by the time the record date arrives, the ex-dividend cutoff has usually already done the sorting. If you bought before the ex-dividend date, you'll be treated as eligible. If you bought on or after it, you won't.

A good way to think of the record date is like the guest list being printed. The ex-dividend date is when the bouncer starts enforcing it.

Payment Date: The Mailman Shows Up (Finally)

Payment date is when the cash is distributed. Sometimes it's the next day. Sometimes it's weeks later. That lag creates a psychological trap: investors watch the calendar, waiting for money to arrive, as if the payment date itself is the key that unlocks the dividend.

But the market is forward-looking, and the entitlement is already settled by the cutoff. By the time payment date arrives, you're mostly just waiting for the operational pipes to deliver what was already promised.

One more practical truth: payment date is when your brokerage may display the cash as "pending" and then "available." If you're counting on that money to pay bills, you want to know your broker's typical posting time and whether the dividend arrives pre-market, intraday, or after-hours. The *company* pays on the payment date; your *experience* of getting paid is filtered through the broker's processing.

For long-term investing, that's a footnote. For someone treating dividends like a paycheck, it's the difference between calm and chaos.

A Simple Timeline Example (With Real-World Timing Logic)

Let's use a clean example where the record date falls on a normal business day.

- Monday: company declares a dividend.
- Two weeks later (Monday): record date.
- Around that same time: ex-dividend date is set based on exchange rules and settlement conventions; for many U.S. stocks it will be on the record date (or one business day before if the record date isn't a business day). (investor.gov)
- Next day (or later): payment date.

What you do with that information:

- Want the dividend? Buy *before* the ex-dividend date.
- Want to free up capital but still collect this dividend? You can generally sell *on* the ex-dividend date and still receive it.

This is also why "buy the day before the payment date" is usually a losing plan. By then, the shares have already started trading without that dividend attached.

The Price Drop: Why the Dividend Isn't ``Extra"

A dividend feels like a reward. Emotionally, it feels like a company handing you a tip for being loyal.

Financially, it's closer to this: the company takes a bucket of cash that used to belong to *all shareholders collectively* and pours it into *individual shareholder pockets*. The business is now holding less cash, so the equity is worth less, all else equal. That's why the stock price tends to adjust downward on the ex-dividend date by about the dividend amount. (investopedia.com)

This matters because it prevents a dangerous misunderstanding: "I'll buy right before the dividend, collect the cash, and then sell. Easy profit."

That strategy even has a name: dividend capture. And it's one of those ideas that survives because it *sounds* like it should work. In efficient markets, it's usually neutralized by the price adjustment, transaction costs, taxes, bid-ask spreads, and the fact that you're trying to be clever in a stadium full of professionals and algorithms. (investopedia.com)

Dividends are powerful, but not because they're a loophole. They're powerful because they can turn a portfolio into a *cash-flow machine* that doesn't require you to sell shares to generate spendable money.

The "Big Dividend" Exception: When the Calendar Changes

Most dividends are small relative to the stock price. But sometimes a company pays a special dividend so large that the normal ex-dividend timing would create chaos.

U.S. rules draw a bright line at "significant" distributions: if a dividend is *25% or more of the stock's value*, special rules apply and the ex-dividend date is typically deferred until *one business day after the dividend is paid.* (investor.gov)

This is not trivia. It changes who gets paid if shares trade hands between the record date and the payment date.

FINRA Rule 11140 lays out the distinction clearly:

- For distributions *less than 25%* of the security's value, the ex-dividend date is generally tied to the record date (with adjustments for non-delivery days). (finra.org)
- For distributions *25% or greater*, the ex-dividend date is generally the *first business day following the payable (payment) date.* (finra.org)

Why would regulators do this? Because a huge dividend behaves less like a routine cash payout and more like a partial return of capital. If a stock is about to pay out a quarter of its value in cash, buyers and sellers need a trading convention that makes the entitlement unambiguous and prevents accidental "gotchas" in the middle of settlement and delivery mechanics.

If you ever see a headline about a "special dividend," don't assume the usual ex-date logic applies. Look it up. The calendar may be shifted specifically to keep the market orderly.

A Second Confusion Trap: ``But I Owned It on the Record Date!"

People sometimes say, "I owned the stock on the record date, so I should get the dividend." That statement sounds reasonable, and in a purely literal sense it's what the company announced.

But the market's trading rule is built to make entitlement tradable without disputes: shares trade either *with* the dividend attached (cum-dividend) or *without* it (ex-dividend). Once the stock goes ex-dividend, the seller keeps (or owes) the dividend based on those rules.

So if someone buys on the ex-dividend date and then notices they "owned it on the record date" (especially when the record date and ex-dividend date coincide), they can feel cheated. Nothing went wrong. They simply bought the seat after the show started.

This is also why brokerage confirmations and dividend calendars usually emphasize ex-dividend dates, not record dates. For investors, the ex-date is the usable decision point.

Dividends and Options: The Quiet Calendar That Moves Prices

Even if you never touch an options chain, options traders live and die by dividend dates because expected dividends can affect option pricing and early exercise decisions (especially for calls). When you *do* use options to manufacture income—covered calls, collars, cash-secured puts—the ex-dividend date becomes a practical risk-management date.

For example, when you sell a covered call on a dividend-paying stock, you are effectively renting out some of your upside. If the call is in-the-money as the ex-dividend date approaches, there can be an incentive for the call holder to exercise early to capture the dividend. That can cause you to lose the shares *right before* the ex-date and miss the payment.

This isn't a reason to avoid options. It's a reason to respect the calendar. Dividends are not just cash flow; they're a scheduling constraint that can interact with your positions in ways that feel "unfair" if you didn't know the rules existed.

If you like getting paid, treat the ex-dividend date the way you treat an earnings date: a known event that can change behavior around your position.

The Practical Way to Use the Calendar (Without Becoming a Spreadsheet Person)

You don't need to memorize settlement mechanics or FINRA rule numbers to invest well. You need a simple operating rule:

> *If you want the next dividend, you must own the shares before the ex-dividend date.*

Then add three habits that keep you from making the classic mistakes:

- *Look up the ex-dividend date first,* not the payment date. The payment date is when you feel happy; the ex-date is when you make decisions.

- *Expect the price to adjust.* If you "earned" $0.50 in dividend, don't be shocked if the stock is about $0.50 lower the next morning, all else equal.
- *Treat special dividends as special.* If the payout is unusually large (think: a big one-time distribution), confirm whether the ex-date is deferred until after payment under the 25% rule. (investor.gov)

Actionable Takeaway

Before placing any trade in a dividend-paying stock (especially if you're doing it for income), do this two-minute check:

1. Find the stock's *ex-dividend date.*
2. If you want the dividend, make sure your purchase happens *before* that date.
3. If you're running options on the position, mark the ex-dividend date on your calendar as a "do not ignore" day—because the market won't ignore it for you.

3.2 Safety First: Spotting Yield Traps

A dividend yield can look like a warm, glowing "SALE" sign in a storefront window. Ten percent! Twelve percent! Fifteen! Your brain does the instant math: "If I put $100,000 here, I'll get $15,000 a year." It feels concrete. It feels adult. It feels like you found the cheat code everyone else missed.

Then the company cuts the dividend, the stock drops again, and you realize the "SALE" sign was taped to the window of a building that was already on fire.

That's the yield trap in one sentence: a very high dividend yield is often the *symptom* of a problem, not the prize for being clever. The market is not in the habit of offering double-digit cash yields for free. When a yield spikes, it's frequently because the price is falling faster than management can write reassuring press releases.

Why Yield Spikes Right Before Things Get Ugly

Dividend yield is just a fraction:

$$\text{Dividend Yield} = \frac{\text{Annual Dividend per Share}}{\text{Stock Price}}$$

So the yield can jump in two ways:

- the dividend goes up (rarely explosive, usually incremental), or
- the price goes down (often fast, often for a reason).

When people get trapped, it's usually the second one. The dividend hasn't improved; the business's prospects have deteriorated. The market reprices the stock lower, but the dividend amount on the brokerage screen is still the old number, like a restaurant menu that hasn't been updated since inflation hit.

That stale dividend number creates a dangerous optical illusion: a yield that looks generous precisely when the dividend is becoming least reliable.

A Quick Reality Check: Dividends Are Paid With Cash, Not Confidence

A company can *declare* a dividend with a board vote and a PDF on its website. It can only *pay* the dividend with cash. And cash has only a few sources:

- cash generated by the business (operations),
- cash borrowed (debt),
- cash raised by selling assets or issuing shares (which is often a warning sign when used to fund payouts).

If you want to avoid yield traps, you don't start by asking "What's the yield?" You start by asking a more uncomfortable question:

> *Where is the money coming from, and how hard is it going to be to keep it coming?*

That question leads straight to the most useful concept in dividend safety: the payout ratio.

The Payout Ratio: The Dividend's Fuel Gauge

A payout ratio is a simple idea: *what percentage of the company's profit (or cash) is being sent out the door as dividends?*

There are two payout ratios worth caring about. One is popular because it's easy. The other is powerful because it's real.

1) Earnings Payout Ratio (Easy, Common, Sometimes Misleading)

The classic payout ratio uses earnings per share (EPS):

$$\text{Earnings Payout Ratio} = \frac{\text{Dividends}}{\text{Net Income}}$$

If a company earns $2.00 per share and pays $1.00 per share in dividends, the payout ratio is 50%. That sounds reasonable: half reinvested, half paid out.

The problem is that earnings are an accounting measure. Accounting matters, but it doesn't always answer the question, "Do they have the cash?"

A company can report earnings while simultaneously:

- spending enormous amounts on capital expenditures (capex),
- getting squeezed by working capital swings (inventory, receivables),
- paying high interest expense,
- facing one-time charges that make EPS look "temporarily" ugly.

Those realities don't care about tidy narratives. They care about the bank account.

2) Free Cash Flow Payout Ratio (Harder, More Honest)

Free cash flow (FCF) is the cash left over after the business pays its operating bills and invests to keep the machine running (capex). A common safety lens is:

$$\text{FCF Payout Ratio} = \frac{\text{Dividends Paid}}{\text{Free Cash Flow}}$$

This is the dividend's *fuel gauge*. If the company is paying out 90% of free cash flow, it's driving with the needle hovering over empty. That doesn't guarantee a cut tomorrow morning, but it makes the dividend sensitive to bad luck: a recession, a spike in costs, a capex cycle, a regulatory change, a single ugly quarter.

Many investor-education sources emphasize that free cash flow provides a more realistic picture of dividend sustainability than earnings-based measures because it focuses on actual cash generated after capital spending. (help.dividenddata.com)

The Traffic-Light Test (A Practical, Not-Perfect System)

No metric can predict dividend cuts with certainty. Management can choose to cut even when they *could* pay, and sometimes they maintain dividends longer than seems rational because they know investors treat a cut like a scarlet letter.

Still, you want a fast way to sort opportunities into "worth deeper research" versus "this is probably a trap."

Here's a simple traffic-light system based primarily on free cash flow payout ratio, with a reality-check glance at earnings:

- *Green (generally healthy)*: FCF payout ratio under about 60%
  Plenty of room for bad quarters, reinvestment, and balance-sheet support.
- *Yellow (watch closely)*: roughly 60%–90%
  The dividend can be fine, but it's dependent on execution and stable conditions.
- *Red (danger zone)*: consistently over 90%
  A small stumble can force a cut, borrowing, or asset sales.

This is deliberately blunt. It's designed to keep you from being seduced by a headline yield while ignoring the math underneath.

Two important caveats keep this system from becoming a trap of its own:

- Different industries have different "normal" cash patterns. Utilities, pipelines, and other capital-heavy businesses can have stable cash flows and high payout ratios for long stretches. Tech companies often have lower payout ratios because they retain more for growth.
- Some sectors require different denominators. REITs, for example, are often evaluated using funds from operations (FFO) rather than EPS because depreciation makes earnings look artificially low even when cash flow is healthy.

The goal isn't to become a sector accountant. The goal is to avoid the universal mistake: treating yield like a safety score.

Case Study #1: Intel (When a Familiar Name Still Cuts Hard)

Dividend cuts don't just happen to obscure companies with questionable balance sheets and a logo you've never seen. They happen to household names when the business model is under pressure and cash needs to be redirected.

In February 2023, Intel announced a 66% reduction in its quarterly dividend, cutting it from $0.365 to $0.125 per share. (apnews.com)

The detail that matters isn't the drama of the announcement. It's the logic behind it: a dividend is a capital-allocation decision. If the company believes the best use of cash is reinvestment, restructuring, debt reduction, or simply survival, the dividend becomes optional. Painful, reputation-damaging, but optional.

A yield trap often looks like this in real time:

- the stock price falls (yield rises),
- investors "lock in" the high yield by buying,
- the dividend is cut,
- the investor now owns a weaker stock with a lower dividend and a bruised market perception.

By the time the cut is announced, the market has often been signaling trouble for months in the only language it speaks fluently: price.

Case Study #2: AT&T (When ``Income Stock" Becomes ``Corporate Action Stock")

Sometimes the dividend is cut not because the company is collapsing, but because the company is changing shape.

AT&T, long held as a classic income stock, reduced its dividend as part of the WarnerMedia spinoff/Discovery transaction. The annual dividend was reset from $2.08 to $1.11 per share. (axios.com)

If you bought AT&T purely for the yield without understanding the business events driving the cash needs, you weren't just taking "dividend risk." You were taking *corporate action risk*—the risk that management will decide the old payout no longer fits the new company.

This is a different flavor of yield trap: not a slow bleed, but a strategic reset. The lesson is the same: dividends are not contracts carved in stone. They are policies, and policies can change when the company's priorities change.

The ``Dividend Isn't Safe" Checklist You Can Run in 10 Minutes

A good checklist doesn't try to be clever. It tries to be hard to fool.

When a yield looks unusually high, run these checks before you get emotionally attached to the number:

1) Did the Yield Rise Because the Price Fell?

Look at a one-year chart. If the price is down 30% and the yield is up dramatically, assume the market is warning you about something.

That "something" might be temporary. It might even be an opportunity. But don't confuse "opportunity" with "income you can safely spend."

2) Is Free Cash Flow Covering the Dividend?

If free cash flow is negative or inconsistent, the dividend has to be funded another way. Sometimes that's fine for a short period. Sometimes it's a countdown clock.

Even without building a spreadsheet, you can usually find "free cash flow" and "dividends paid" in financial summaries, earnings releases, or reputable company analysis platforms. The exact sources vary, but the logic doesn't: cash in, cash out.

3) Is the Company Using Debt to Fund Dividends?

Borrowing to pay a dividend is like using a credit card to pay your mortgage. It can get you through a rough patch, but it's not a long-term lifestyle.

A company with rising leverage and a high payout ratio is often juggling. The dividend becomes the easiest ball to drop if lenders, ratings agencies, or interest rates get unfriendly.

4) Is Management Signaling a "New Priority"?

Listen to what executives emphasize:

- "Balance sheet strength"
- "Investing in transformation"
- "Strategic flexibility"

Those phrases can be sensible and shareholder-friendly. They can also be early warnings that dividend growth is about to pause, or the dividend itself is about to shrink.

When you see those signals *and* a stretched payout ratio, you're not looking at a stable income machine. You're looking at a management team preparing the market for change.

A Sneaky Side Effect: Yield-Chasing Can Turn Your Dividends Into Higher-Taxed Income

This is where the calendar of dividends quietly matters again, but in a different way.

In the U.S., dividends can be "qualified" (taxed at preferential long-term capital gains rates) or "non-qualified" (taxed like ordinary income). One key requirement for qualified dividend treatment is a holding period rule: for common stock, you generally must hold the stock for more than 60 days during the 121-day period that begins 60 days before the ex-dividend date. (irs.gov)

If you buy a stock right before the ex-dividend date and flip it soon after—classic yield-chasing behavior—you can easily fail the holding period test and end up with dividends taxed at higher ordinary income rates, even if your broker reports them in the "dividends" bucket.

There's a deeper point here than tax trivia: strategies that revolve around grabbing a dividend and running tend to be fragile. The market often neutralizes the dividend via price adjustment, and the tax code can reduce what you keep. A yield trap can hurt you in multiple dimensions at once: price, income, and after-tax reality.

What ``Dividend Safety" Actually Means (And What It Doesn't)

Dividend safety doesn't mean the company will never cut. It means the company has room to keep paying through normal adversity.

Think of a safe dividend like a well-built bridge. It's not guaranteed to last forever. But it has:

- structural support (cash generation),
- maintenance budget (reasonable payout ratio),
- load capacity (ability to handle a downturn without snapping).

A yield trap is a bridge that looks impressive from far away because it's painted bright gold. Up close, you see missing bolts.

The mistake most investors make is focusing on *current income* while ignoring *future flexibility*. Companies cut dividends when flexibility disappears.

Actionable Takeaway

Before buying any stock because the yield looks "too good to pass up," write down the answers to these two questions:

1. *Is the dividend covered by free cash flow over the last year (or last few years), or is the company stretching?*
2. *Did the yield rise because the business improved, or because the stock price fell and the old dividend number hasn't caught up with reality?*

If you can't answer those in plain language, treat the high yield as a warning label, not an invitation.

3.3 The Great Debate: Growth vs. High Yield

Picture two investors sitting at the same kitchen table, staring at the same brokerage screen, both saying the same sentence: "I want income."

Investor A means: "I want *money this month.*" Rent doesn't care about your total return chart. Groceries don't accept unrealized gains. The only thing that feels real is cash landing in the account on schedule.

Investor B means: "I want *money for the rest of my life.*" They're not trying to maximize this year's payout; they're trying to build a dividend stream that keeps getting raises long after their salary stops.

Same word. Totally different job.

That difference is why the argument between *high yield* and *dividend growth* never dies. It's not really a fight about stocks. It's a fight about *time*.

Two Income Personalities Hiding Inside One Portfolio

High-yield investing is the income strategy with a loud engine. It hands you big checks early, often from businesses designed to distribute cash: utilities, telecoms, pipelines, certain REITs, closed-end funds, covered-call ETFs, and other yield-oriented vehicles. It's the tortoise strategy in a cash-flow sense: steady, deliberate, and unapologetically focused on the paycheck.

Dividend-growth investing is the income strategy with a quiet engine. It starts smaller, sometimes almost insultingly small. A 1.5% yield doesn't feel like "income" when you're trying to cover bills. But the point is not the first year. The point is the *raises*. A company that can increase its dividend 8% per year for a long time has a way of making the early years look like the prologue.

The psychological difference is huge:

- "Relax. The money is already here."
- "Be patient. The money is being built."

Neither is morally superior. They're tools. And like all tools, the wrong one at the wrong time can hurt you.

The Hidden Trade: Immediate Cash vs. Inflation-Proof Cash

The most important enemy of income isn't volatility. It's *erosion*.

A flat dividend can feel safe because it keeps coming. But if the dividend doesn't rise, your purchasing power quietly shrinks. Ten years later, you're still receiving the same dollar amount, but your bills have learned new tricks.

That's the brutal math of retirement income: you don't just need cash flow; you need *cash flow that can keep up with life*.

High-yield portfolios often have slower dividend growth. Sometimes that's by design: mature businesses with limited reinvestment opportunities distribute more today but grow less tomorrow.

Dividend-growth portfolios often have lower starting yields because the companies are still reinvesting heavily. They pay less now because they're trying to become more valuable later. When that reinvestment succeeds, dividend increases can become part of the company's identity—not a marketing gimmick, but a discipline.

The real debate is whether you prefer:

- a big check that may not grow much, or
- a smaller check that keeps getting raises.

Most investors don't need to pick one forever. They need to pick what fits *this stage* of their financial life.

A Grounding Principle (That Sounds Cold, But Helps You Think Clearly)

In an idealized world with no taxes, no transaction costs, and perfectly rational investors, dividend policy shouldn't change the value of a company. If a company pays you a dividend, the stock price drops by roughly that amount; if it keeps the cash, the value stays in the business. Same pie, different slices.

Real markets are not that idealized world. But the principle is still useful because it punctures a common fantasy: that dividends are "extra."

Dividends are *one way* a business delivers shareholder return. They can also be a management discipline (a self-imposed rule to avoid wasteful spending), a signal (confidence), or a shareholder-friendly choice in a mature business. But they aren't a magic portal that produces wealth without tradeoffs.

So when you compare high yield vs. dividend growth, don't compare them like one is "return" and the other is "not return." Compare them as *two packaging styles* for how return shows up in your life.

High Yield: When the Checks Need to Be Big *Now*

High yield shines when time is short or income needs are immediate. That could mean:

- you're retired (or semi-retired) and want the portfolio to carry part of the household budget,
- you're building an "income floor" that reduces reliance on selling assets,
- you prefer emotional stability: watching cash arrive helps you stay invested when markets get ugly.

There's also a practical advantage: high yield can reduce the temptation to sell. A portfolio that throws off meaningful cash can feel like it's doing its job even when prices bounce around.

For many people, that's not a behavioral footnote—it's the difference between sticking to a plan and panicking.

But high yield has two repeating hazards:

Hazard 1: The "Yield Trap" Gravity

The higher the yield, the harder you must interrogate the source. If the yield is high because the price fell, you may be staring at a business whose cash flows are under pressure. That pressure eventually shows up as a dividend freeze or cut, which is like losing a paycheck and taking a capital loss at the same time.

A high yield is not automatically dangerous. It's just a claim that requires evidence.

Hazard 2: Limited Dividend Growth

Many high-yield businesses can't raise the dividend quickly without starving their own future. If you pay out most of your cash, there's less left to reinvest, adapt, and grow.

That doesn't mean you should avoid them. It means you should be honest about what you bought: a *cash-distribution machine*, not necessarily a compounding machine.

High yield is often excellent at funding today. It can be mediocre at funding *tomorrow's cost of living* unless you reinvest a portion, add new capital, or supplement with other income strategies.

Dividend Growth: When You Want Raises More Than You Want a Big First Paycheck

Dividend-growth investing is built on a simple, almost boring preference: companies that can keep increasing dividends tend to be companies with durable cash generation and decent capital allocation habits. Not always. Not perfectly. But often enough that the market has turned "dividend growers" into a recognizable category.

A dividend growth strategy asks a different question than high yield. Instead of "How much does it pay today?" it asks:

> *How likely is this business to be stronger in ten years, and how likely is management to share that strength with me?*

The advantage is not merely higher future income. It's also a different relationship with inflation. If your dividend grows, your portfolio is attempting to hand you a raise each year. That can be emotionally and financially powerful, especially later in life when your ability to "just work more" is limited.

But dividend growth has its own hazards:

Hazard 1: The Patience Tax

A low starting yield can feel like you're doing all the work and getting none of the reward. If you need income soon, dividend growth can be the wrong tool. You can't pay the electric bill with "future compounding."

Hazard 2: Overpaying for "Quality"

The market loves a great story. "High-quality dividend growers" can become crowded trades, where investors pay a premium for stability and consistency. Paying too much can reduce returns for years, even if the business performs well.

Dividend growth works best when you remember you're still buying a stock, not a certificate of virtue.

The Bridge Concept: Yield on Cost (The Number That Explains the Argument)

If you've ever heard someone brag that they own a stock yielding "12%", and then you look it up and it yields "2.5%", you've met *yield on cost*.

Yield on cost (*YOC*) is your annual dividend divided by the price you originally paid. It's a personal metric, not a market metric:

$$\text{Yield on Cost} = \frac{\text{Current Annual Dividend}}{\text{Your Purchase Price}}$$

Why do people love this number? Because it captures the emotional reward of dividend growth. If you bought a dividend grower years ago and it kept raising the payout, your cash income on that original investment can become surprisingly large.

Here's what yield on cost is *good* for:

- reminding you that a modest yield today can become a meaningful income stream over time,
- reinforcing the idea that dividend growth is a long game.

Here's what yield on cost is *not* good for:

- making new decisions.

Markets don't care what you paid. Your next decision should be based on what the stock offers *from here*: current yield, growth prospects, valuation, risk. YOC can be a satisfying trophy, but it's not a steering wheel.

Still, YOC is the cleanest way to explain why these two strategies argue so much. High yield starts strong and often grows slowly. Dividend growth starts small and (if successful) accelerates.

The strategies are not enemies. They are time-shifted versions of "income."

A Concrete Example (With Numbers That Behave Like Real Life)

Assume you have $100,000 to allocate.

High Yield Portfolio: 7% yield, dividend growth 1% per year (slow growth).
Dividend Growth Portfolio: 2.5% yield, dividend growth 10% per year (fast growth).

Year 1 income:

- High yield: $7,000
- Dividend growth: $2,500

If you need $5,000 to support your lifestyle, only one of these strategies is doing the job without selling shares. That's not a theoretical insight; that's a monthly-budget fact.

But fast-forward. The dividend-growth portfolio's income grows at 10% per year. After 10 years, its annual income is roughly:

$$\$2{,}500 \times (1.10)^{10} \approx \$6{,}484$$

Meanwhile the high-yield portfolio grows slowly:

$$\$7{,}000 \times (1.01)^{10} \approx \$7{,}732$$

At 10 years, high yield is still ahead. At 15 years, the growth portfolio is roughly:

$$\$2{,}500 \times (1.10)^{15} \approx \$10{,}443$$

High yield at 15 years:

$$\$7{,}000 \times (1.01)^{15} \approx \$8{,}123$$

Now the argument flips. The "small yield" strategy becomes the bigger paycheck. Not because it was secretly better at the start, but because time finally did what time does: it rewarded the strategy that was designed for compounding.

A Quick Comparison Table (So You Can Match Strategy to Life)

	High Yield Focus	*Dividend Growth Focus*
Primary goal	Maximum cash flow today	Rising cash flow over time
Best for	Near-term income needs; building an income floor	Long horizons; inflation-fighting income
Common risk	Yield traps; dividend cuts; slow growth	Overpaying; slow early income; patience required
What success feels like	"The portfolio pays my bills."	"My income gets raises."
Portfolio behavior	Often steadier income, potentially less price upside	Often stronger long-run income growth, variable valuation cycles

This is not a personality test. It's a job description. Pick the job first. Then hire the strategy.

Where Options Quietly Change the Debate

Options don't replace dividends, but they can change what you need from them.

If you know how to responsibly generate premium income (for example, covered calls on positions you're happy to hold, or cash-secured puts as a disciplined entry method), you can sometimes afford to own lower-yielding, higher-quality companies and still produce meaningful cash flow in the early years.

That matters because one of the biggest problems with dividend growth is the "thin paycheck" phase. Options premium can act as a bridge:

- Dividend growth supplies the long-run raises.
- Options premium supplies supplemental cash today (with its own risks and tradeoffs).

The point isn't to force options into every portfolio. The point is to recognize you're not stuck choosing between "income now" and "income later" as a binary. You can build a system where different cash-flow sources carry different time horizons.

The Crossover Point (When Growth Finally Wins the Paycheck Contest)

The whole debate becomes crystal clear when you draw it.

The high-yield income line is high and relatively flat. The dividend-growth income line starts lower, then curves upward, and eventually crosses the high-yield line. That crossing is not guaranteed in real investing (dividend growth rates vary; companies stumble; portfolios change). But as a mental model, it's a remarkably honest picture of the trade.

Below is a stylized chart using the same kind of assumptions discussed above: high yield starts strong but grows slowly; dividend growth starts small but compounds faster. The key feature is the *crossover point*—the year when the growth strategy's annual income surpasses the high-yield strategy.

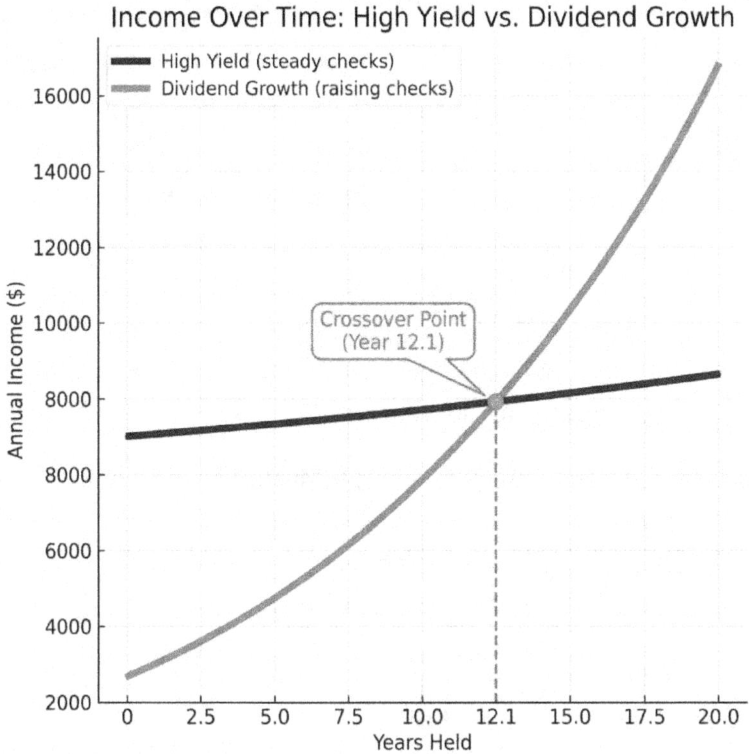

The chart doesn't declare a winner. It exposes the *price of admission* for each approach:

- High yield asks you to accept slower income growth later in exchange for bigger checks now.
- Dividend growth asks you to accept smaller checks now in exchange for faster growth later.

If you're 35, the crossover point is your friend. If you're 75 and funding your life from the portfolio, the crossover point may be irrelevant. A strategy that "wins" in year 14 is not comforting if you need the income in year 1.

The Most Useful Way to Stop Arguing: Build a Barbell

Some investors find peace by building a barbell:

- One side of the barbell focuses on reliable current income (but with strict safety standards).
- The other side focuses on dividend growth (so the income stream can get raises).

The portfolio then behaves like a household with two earners:

- one earns a steady salary,
- the other is on a fast career track but starts lower.

You're not betting your entire future on one kind of paycheck. You're diversifying across *time*.

That structure also gives you a practical rule for reinvesting: when the income side is doing its job, excess cash can be directed toward growth—especially when growth stocks are temporarily out of favor and priced more reasonably.

Actionable Takeaway

Write down your honest time horizon for income in one sentence, then choose your strategy mix accordingly:

> *"I need $___ per year from my portfolio starting in (month/year), and I want that income to grow by roughly $___ per year to keep up with life."*

If the start date is soon, weight toward safe current income and treat dividend growth as a supporting actor. If the start date is far away, weight toward dividend growth and use current yield as a bonus, not the mission. The debate stops being philosophical the moment you attach it to a calendar and a number.

3.4 To DRIP or Not to DRIP

The first time dividend income starts to feel "real," it's rarely because the numbers are huge. It's because the money shows up while you weren't watching. You wake up, open your account, and there it is: cash from companies you didn't invoice, didn't negotiate with, and didn't have to beg for a raise from. It's delightfully lazy income.

Then comes the fork in the road that quietly shapes the rest of your investing life: do you automatically roll that cash back into the same stocks (DRIP), or do you collect it into a pile and decide what to do with it later?

That decision sounds like a minor settings toggle. It isn't. It's an operating system choice.

What a DRIP Actually Does (And Why It Feels So Good)

A dividend reinvestment plan—a DRIP—simply takes the cash dividend you were owed and uses it to buy more shares (often including fractional shares). Your share count creeps higher without you doing anything. The next dividend is paid on a slightly larger number of shares. The next reinvestment buys a bit more. Over years, it can snowball into a meaningful compounding engine. Britannica's investor education coverage captures the modern reality: many brokers offer automatic reinvestment, often including fractional shares, and investors can use reinvestment to make dividend-paying positions part of a growth strategy. (britannica.com)

The emotional appeal is hard to overstate. DRIP turns the investing experience from "I hope I'm right" into "I'm accumulating." Even in flat markets, your share count can rise. That rising share count is a small psychological anchor in a world that otherwise loves to wobble.

But DRIP isn't "more disciplined investing" by default. It's *automatic investing*. Automation is only disciplined when the default behavior is the behavior you would have chosen on purpose.

Two Flavors of DRIP: Broker DRIP vs. Company DRIP (They Are Not the Same Creature)

People use "DRIP" as one word, but it often means two different mechanisms.

Brokerage dividend reinvestment is the common modern version: you check a box in your brokerage settings and the broker reinvests dividends back into the paying security (or sometimes into a designated choice, depending on the platform).

Company-sponsored DRIPs (often tied to direct stock purchase plans) are older-school and still exist for some issuers. The

SEC's Investor.gov notes that direct stock plans and issuer DRIPs can have their own rules and fees, and importantly, they often execute trades at set times and at an average market price rather than letting you buy at a specific moment. (investor.gov)

That "set times / average price" detail sounds harmless until you've lived through a volatile week. If the market whipsaws and you expected to reinvest on a dip, the plan may purchase on its schedule anyway. That isn't automatically bad. It's just different. If you value control and simplicity, broker reinvestment is typically cleaner. If you value direct ownership through an issuer plan, you accept the plan's mechanics.

Either way, the core question is the same: should this dividend cash be forced back into *the same* asset immediately?

DRIP's Superpower Is Also Its Weakness: It Never Asks ``Is This a Good Price?"

A DRIP is a loyal employee who never questions management. It buys the stock because the policy says so, not because the price is attractive.

That can be wonderful when:

- you're accumulating for decades,
- you're buying quality businesses you'd happily add to at most prices,
- the dividend payer is a diversified fund where "same asset" still means "broad basket."

It can be actively harmful when:

- the stock has become meaningfully overvalued,
- the business has deteriorated (but the dividend hasn't been cut *yet*),
- your position has grown too large and DRIP keeps feeding the concentration.

That last point is the silent portfolio-killer. DRIP doesn't just compound your wealth. It compounds your *weights*. If one holding becomes a winner (or simply stays high-priced while others lag), reinvestment keeps buying more of the thing you already have plenty of. Over time, you can unintentionally turn a diversified income portfolio into a handful of oversized bets.

If you've ever said, "How did this become 18% of my account?", DRIP is often part of the answer.

The Tax Surprise: Reinvested Dividends Are Still Taxable (In Taxable Accounts)

Many investors learn this the hard way: automatic reinvestment does not make dividends disappear for tax purposes. The IRS treats reinvested dividends as income in the year paid, even if you never touch the cash.

The IRS FAQ on reinvested dividends is blunt: if reinvested dividends buy shares at fair market value, you still report the dividends as income; and if the DRIP lets you buy shares below fair market value, you generally must also report as dividend income the fair market value of the additional stock received on the payment date. (irs.gov)

Publication 550 reinforces the same idea and adds a practical basis detail: if you receive stock through a DRIP at a discount, you include the discount in income and your basis is the full fair market value on the dividend payment date. (irs.gov)

This matters operationally for two reasons:

- *Cash-flow planning*: you may owe taxes even though you never "took" the dividend in cash. DRIP can quietly create a tax bill without creating spendable dollars.
- *Recordkeeping*: DRIP creates many tiny tax lots. Your broker will usually track basis, but the more complex the history (transfers, multiple accounts, older holdings), the more you want to ensure your records are clean.

In retirement accounts (IRAs, 401(k)s), this issue largely goes away because current taxes are deferred (traditional) or not owed on qualified distributions (Roth). In taxable accounts, it's real.

DRIP is not a tax strategy. It's a compounding strategy. Treat it accordingly.

A Second Tax Detail Most People Miss: DRIP Can Interact With ``Qualified Dividend" Treatment

Dividend taxes aren't one flat rate. In the U.S., "qualified dividends" can be taxed at preferential long-term capital gains rates, but you must meet holding-period rules tied to the ex-dividend date.

IRS Publication 550 lays out the common-stock holding-period requirement: you must hold the stock for more than 60 days during the 121-day period that begins 60 days before the ex-dividend date. (irs.gov)

This doesn't mean DRIP is dangerous. In fact, DRIP tends to push you toward longer holding periods, which is generally compatible with qualified dividend treatment. The bigger risk shows up when investors toggle behaviors:

- buying right before dividends,
- selling soon after,
- rotating in and out of "high yield this month" ideas.

If your dividend habits become short-term and twitchy, the tax code becomes less friendly. The irony is almost comedic: chasing income can make your income taxed more like ordinary wages.

When DRIP Is the Right Default

DRIP is a great default when you want the portfolio to behave like a self-replicating organism: take in nutrients (dividends), convert them into more productive cells (shares), repeat.

Practically, DRIP tends to make sense when:

3.4.1 You're in accumulation mode and don't need the income to live

If you're not spending the dividends, then every dividend is a decision: reinvest or redeploy. DRIP simply chooses "reinvest" without asking permission. For an investor with a long runway, that can be exactly the behavior you'd choose anyway.

The danger in accumulation is usually not "missing the perfect buying moment." It's forgetting to buy at all. DRIP solves that problem ruthlessly.

3.4.2 The holding is broad or rules-based (funds, diversified ETFs)

Reinvesting into a single company at any price can be questionable. Reinvesting into a diversified vehicle is usually less so because you're effectively adding to a basket. You're not doubling down on one management team's judgment.

3.4.3 You're deliberately dollar-cost averaging into a business you understand

DRIP works best when the underlying holding is something you would keep buying manually. If you wouldn't buy more at today's price with fresh cash, you should hesitate to buy more with dividend cash just because the default setting says so.

A DRIP should feel like a policy you endorse, not a habit you forgot you had.

When DRIP Is the Wrong Default (Even If You Love Compounding)

There are moments when turning off DRIP is not an act of laziness; it's an act of adult supervision.

3.4.4 When you're building (or protecting) a cash buffer

The most underrated income strategy is having enough cash on hand that you don't have to sell assets at the worst possible time.

A *cash buffer* is simply a pool of safe, liquid money (often in a high-yield savings account, money market fund, or short-term Treasuries) designed to cover expenses when markets misbehave. If you're living partly on portfolio income, a cash buffer can be the difference between calm and panic.

In that context, reinvesting every dividend automatically is like insisting every paycheck must go straight into buying company stock even when your checking account is low. It's aggressively "optimized" right up until life happens.

3.4.5 When valuation is obviously stretched

You don't need a PhD in finance to notice when a stock you own has become priced for perfection. If your dividend cash is automatically buying at that moment, you may be compounding into a future disappointment.

This is not a call to time markets with surgical precision. It's permission to use your eyes.

3.4.6 When the position is becoming too large

Concentration risk is seductive because it grows during good times. DRIP helps it grow quietly.

A practical rule some income investors adopt: if any single holding drifts above a chosen percentage of the portfolio (say

5% or 7% or 10%), they stop reinvesting dividends into it. Not forever, just until weights normalize.

Turning off DRIP in an oversized position is a gentle way to rebalance without selling.

3.4.7 When the dividend's safety has become questionable

Sometimes the market's message is subtle: the yield rises, the stock slides, the payout ratio stretches, management starts using phrases like "strategic flexibility." That's the moment when automatic reinvestment can become automatic denial.

If you suspect a yield trap forming, the last thing you want is a robot buying more for you every quarter.

The ``Pool the Dividends" Alternative: Treat Cash Like Ammunition, Not Confetti

Turning off DRIP doesn't mean you stop reinvesting. It means you stop reinvesting *blindly*.

Pooling dividends is simple: dividends land as cash, you let them accumulate, and periodically you deploy them where they'll do the most good. That could mean:

- buying the most undervalued holding on your watchlist,
- adding to a position that's temporarily out of favor,
- funding a new position to improve diversification,
- topping off your cash buffer,
- or simply paying for life.

This approach makes the portfolio feel less like an autopilot and more like a business you run. Dividends become working capital.

The tradeoff is attention. DRIP is automatic and steady. Pooling requires you to make decisions—and decisions are where people get emotional.

If you're the type to freeze up when it's time to buy, pooling can become "cash drag," where money sits idle because you're waiting for a perfect moment that never arrives.

A good compromise is to pool dividends but set a rule like: "When cash reaches $1,000 (or $2,500), I invest it within one week." Rules beat moods.

A Practical Operating Policy That Fits Most Real Lives

Most investors don't need an ideological commitment to DRIP or anti-DRIP. They need an operating policy that adapts to what the portfolio is trying to accomplish right now.

Here's a policy that tends to work well:

3.4.8 If you need the income to live: don't DRIP by default

If dividends are paying bills, automatic reinvestment is fighting your cash-flow needs. Keep it simple: dividends land, you spend what you need, and you reinvest only what's left.

3.4.9 If you don't need the income: DRIP can be the default, with ``off-ramps"

Use DRIP for holdings that are:

- diversified (funds) or truly long-term core positions,
- not overweight,
- and not showing dividend-safety stress.

When any of those conditions fail, turn DRIP off for that holding and pool the cash until you have a better use.

This isn't complicated. It's just treating reinvestment as a decision, not a reflex.

A Note on Mechanics: Timing, Fees, and the ``Why Did It Buy at That Price?" Moment

Even with broker DRIP, reinvestment often happens on a set schedule (commonly the payable date or shortly after the dividend settles). Company direct plans are even more explicitly batch-driven, purchasing at set times and average prices, and may include plan fees. (investor.gov)

So if you ever see a reinvestment price that looks "off," it may simply be the plan's execution method. The right response is not outrage; it's matching the tool to the job:

- If you want set-it-and-forget-it compounding, accept batch execution.
- If you want tactical entry prices, pool dividends and place your own trades.

Also remember that DRIP creates fractional shares. Fractional shares are wonderful for compounding, but they can complicate transfers between brokers and certain corporate actions. It's not a reason to avoid them; it's a reason to know they exist.

Actionable Takeaway

Choose one of these two operating modes for each holding in your income portfolio, and write it down:

- *Auto-Compound Mode* (*DRIP ON*): for diversified funds or core holdings that are not overweight and have no dividend-safety red flags.
- *Owner-Operator Mode* (*DRIP OFF*): dividends flow into cash, first filling a cash buffer (if needed), then being deployed intentionally into the most attractive opportunity in your portfolio (or used for spending).

Then add one hard rule that prevents "accidental concentration": if any single holding exceeds your chosen maximum portfolio weight, turn DRIP off for that holding until it's back in range.

That's it. One toggle, one rule, and you've turned dividends from random droplets into a cash-flow system you can actually run.

Chapter 4

Options Fundamentals for Income

If stocks are the engines of wealth, options are the gears that let you control the speed and direction. Too often, options are sold to the public as lottery tickets-risky bets on the next big explosion. But in the Income Factory, we don't buy lottery tickets; we sell them. In this chapter, we strip away the Wall Street jargon to reveal options for what they really are: insurance contracts. You will learn how to play 'the house,' collecting premiums from speculators and hedgers alike, and how to harness the passage of time to generate reliable cash flow.

4.1 The Language of Options

The first time most people hear the word *option*, they picture a day-trader staring at a screen like it's a slot machine-fast, loud, and vaguely dangerous. That stereotype sticks because the *buyer* of an option really is making a wager: paying cash today for the possibility that something dramatic happens tomorrow. But income investors don't have to live in the buyer's world. The income investor's natural habitat is the other side of the trade: getting paid *up front* for agreeing to a clearly defined deal.

That mindset shift matters because it changes what you root for. Option buyers need movement. Big movement. The kind that makes headlines. Option sellers can get paid for something far more common and far more reliable: *time passing* without catastrophe.

Think of an option contract as a very specific kind of promise, sold in a very standardized package. It is not a vague handshake agreement like, "I might buy your house someday." It's closer to a signed purchase agreement with a price and a deadline—except the buyer has the *choice* to use it, and the seller has the *obligation* to honor it if the buyer chooses to act.

That's the language worth learning: not because options are complicated, but because the names can trick you into thinking they're something else. Once you translate the vocabulary into normal life, most of the fear evaporates.

The contract: what an option *is* (and what it isn't)

An option is a contract on an underlying asset (usually a stock or ETF) that lasts until a specific expiration date and references a specific strike price. Almost everything confusing about options is just people losing track of those four words: *underlying, strike, expiration, premium.*

Here's the part that should make income investors sit up straighter: the premium (the price of the option) is paid *immediately*. Options are one of the few places in markets where getting paid first is normal. You don't mail the premium later. You don't "earn it over time" the way a bond coupon accrues. It hits the account when the trade is opened.

But collecting premium is not "free money" any more than collecting a security deposit from a tenant is free money. The deposit is compensation for risk and responsibility. If the tenant trashes the place, that deposit suddenly has a job to do. Option premium works the same way: it is the market's way of pricing the possibility that you'll be required to do something inconvenient.

That "something inconvenient" is exactly specified. It is never mysterious. It is always written into the contract.

Calls and puts: the two promises

There are only two types of options:

- A *call* is about buying stock.
- A *put* is about selling stock.

If you're the *buyer*, a call gives you the right to *buy* at the strike price; a put gives you the right to *sell* at the strike price.

If you're the *seller*, flip the sentence:

- When you sell a *call*, you may be obligated to *sell* stock at the strike price.
- When you sell a *put*, you may be obligated to *buy* stock at the strike price.

Notice how the seller's version sounds a lot like regular investing. Buying stock at a price you agreed to? Selling stock at a price you agreed to? That's not exotic-it's basically the stock market with deadlines and paperwork.

The real difference is timing: you agree *now* to a trade that might happen *later*. The buyer is paying for flexibility. The seller is being paid for inflexibility.

The four basic positions (and why only two matter for income)

Every option position is one of four building blocks:

- Long Call (buy a call)
- Short Call (sell a call)
- Long Put (buy a put)

- Short Put (sell a put)

The long positions are the buyer's world: limited loss, potentially large gain, and a constant leak of value as time passes. Great if you correctly predict a big move quickly. Brutal if you're merely "kind of right" but late.

The short positions are the seller's world: you start with cash in hand and you're paid for the passage of time. The tradeoff is that you take on an obligation. That obligation can be managed, hedged, rolled, or closed-but it can't be wished away.

For income-oriented strategies, the gravitational center is the short side:

- *Short puts* often behave like getting paid to place a limit order to buy a stock you already want.
- *Short calls* (often covered by shares you own) behave like getting paid to set a "sell if it gets expensive" price tag.

If that sounds too ordinary for something as hyped as options, good. The income approach is ordinary on purpose. It tries to turn options from a prediction contest into a pricing-and-process business.

The landlord and the insurer: two mental models that actually work

Two analogies capture most of what a premium seller is doing.

The landlord. A landlord owns an asset and rents out *usage rights*. The tenant doesn't buy the building; they buy time inside it. The landlord gets paid first (rent), and that rent is compensation for wear-and-tear risk and vacancy risk.

A covered-call seller is doing something similar. You own the shares (the building) and you rent out a very specific right: the right for someone else to buy those shares from you at a fixed price before a fixed date. They pay you rent (the premium) for that right.

The insurer. An insurance company collects small premiums from many people. Most of the time, nothing "happens," and that's exactly what makes the business work. When something does happen, the insurer pays. The insurer survives because it prices risk, sets limits, and never bets the company on a single policy.

A put seller is selling protection against a drop in the underlying. The buyer of a put is buying peace of mind: if the stock falls below the strike, they can sell at the strike. If you sell that put, you're the insurer. You're collecting a premium to shoulder a defined slice of downside risk.

Both analogies point to the same truth: premium selling is a *risk business*. Not a lottery business. You don't need to be right about the future in a cinematic way. You need to be paid appropriately for the risk you agree to carry, and you need to avoid writing policies so large that one storm wipes you out.

Strike price and expiration: the used-car price tag and the "good until" date

If options jargon ever starts to feel like a foreign language, translate it into a used-car listing.

Imagine you're selling your old truck. You write:

> "I will sell this truck for $25,000 any time before March 15."

That's basically a call option contract from the buyer's perspective. The price is the *strike*. The deadline is the *expiration*. If the market suddenly goes crazy and similar trucks sell for $30,000 next week, the buyer loves having that right. If the market softens and trucks sell for $22,000, the buyer ignores it.

Now notice what you'd demand for offering that guarantee. You wouldn't do it for free. You'd want a nonrefundable fee for holding your offer open and taking the risk that prices rise. That fee is the *premium*.

Options simply industrialize this idea. They standardize the contract, list it on an exchange, and make it tradable. But at heart, it's still a price tag plus a "good until" date.

Premium: the money is real, the obligation is real

The premium is the option's price. It's quoted per share, but each standard equity option contract usually controls 100 shares. So a premium of $1.25 means $125 of cash changes hands.

That "times 100" feature is where options get their reputation. People see small numbers and forget the scale. A $0.60 option feels cheap until you remember it's $60 per contract, and you can sell ten contracts in a click. The ease of scaling is a gift when you're disciplined and a trap when you're bored.

For the seller, premium is your paycheck. But it's also a clue. When premium is fat, the market is nervous or uncertain. When premium is skinny, the market is calm-or complacent.

Getting paid well is pleasant. Getting paid well because risk is genuinely higher is not the same thing. The premium tells you what kind of neighborhood you're renting property in.

Intrinsic value vs. extrinsic value: "real equity" and "time & worry"

An option's price has two components:

- *Intrinsic value*: the "already true" part.
- *Extrinsic value*: the "might become true" part.

Intrinsic value is easy to understand because it's basically math you can do on a napkin. A call has intrinsic value when the stock price is above the strike (because you could buy at the strike and immediately sell at the higher market price). A put has intrinsic value when the stock price is below the strike (because you could sell at the strike instead of the lower market price).

Extrinsic value is everything else: time value and implied volatility, bundled together into what you can think of as *time and worry*. Time because there's still runway for the stock to move. Worry because the market is pricing how violent that movement might be.

For income strategies, extrinsic value is the main course. When you sell premium, you are mostly selling *extrinsic value*-you're selling the market's uncertainty. Over time, that uncertainty tends to shrink, like steam leaking from a kettle. The seller gets paid because they're willing to hold the kettle.

A subtle but important emotional point: intrinsic value feels like "fairness." Extrinsic value feels like "someone is overpaying." Beginners often feel guilty collecting extrinsic value, as if it's a trick. It's not. Extrinsic value is the market's price for flexibility. Buyers are not naïve; they're outsourcing risk or expressing a view. You are providing the service.

In-the-money, at-the-money, out-of-the-money: three neighborhoods

Options are constantly described as ITM, ATM, or OTM:

- *In-the-money* (*ITM*): the option has intrinsic value right now.
- *At-the-money* (*ATM*): the strike is near the stock price.
- *Out-of-the-money* (*OTM*): the option has no intrinsic value right now.

To the income seller, these are not status labels-they're risk labels.

An OTM option is mostly extrinsic value. That means you're mainly selling "time and worry." It also means there's a buffer before the obligation becomes urgent.

An ITM option contains a chunk of intrinsic value. When you sell it, you're not just renting out uncertainty; you're agreeing to a deal that is already economically attractive for the buyer. That can still be done intentionally, but it's a different job. You're often taking in less "pure income" per unit of risk and more "stock-like" exposure.

ATM options are where uncertainty is loudest. They often carry the most extrinsic value relative to their price, because a small move can change everything. They can be lucrative to sell, but they demand respect.

A quick translator: how the seller reads common option quotes

Brokers and quotes can make options feel like you're reading a license plate. Under the hood, it's always the same ingredients: underlying, expiration, strike, call/put, premium.

Once translated, a quote is simply a sentence about a deal. For example:

> Sell the XYZ March 21 50 Put for $1.20

In English, that's:

> "Someone pays me $120 today. In exchange, I agree that until March 21, if they choose, I will buy 100 shares of XYZ for $50 per share."

A covered call example:

> Sell the XYZ March 21 55 Call for $0.90 (while owning 100 shares of XYZ)

In English:

> "Someone pays me $90 today. In exchange, until March 21, they can make me sell my 100 shares of XYZ for $55 per share."

That is the seller's world in one breath: *cash now, obligation later, terms clearly written*. If you can translate the quote into that plain-English promise, you're no longer guessing what you own.

The obligation isn't a monster; it's a price you pre-agreed to

Most fear around options comes from the word *obligation*. It sounds like a trap door. In practice, the obligation is just a transaction at a pre-agreed price.

If you sell a put and get assigned, you buy shares at the strike price. That's it. You're not "blown up" by definition. You're not "wrong" by definition. You did the thing you said you were willing to do when you sold the contract.

If you sell a call against shares you already own and get assigned, you sell your shares at the strike price. Again, that's it. You gave someone the right to buy at that price; if the stock soared and they exercise, it means your shares got called away at the price you chose.

The seller's job is to pick strikes that feel like rational prices *even if assignment happens*. That's the discipline that separates income options from adrenaline options. You don't sell a put at a strike that would make you miserable to own. You don't sell a covered call at a strike that would make you furious to sell.

To make this visceral, imagine you're running a little pawn shop. You can write two kinds of tickets:

- "I'll buy your guitar for $400 anytime in the next month" (short put mentality).

- "I'll sell you my guitar for $600 anytime in the next month" (covered call mentality).

If those numbers are sane-numbers you'd accept with a shrug-then the option contract is not a monster. It's a paid offer.

Why time is the seller's quiet advantage

Option buyers are fighting a clock that never stops. Every day that passes with "nothing happening" is a day the buyer paid for and didn't use. That is why options have decay built into their pricing. The buyer isn't just betting on direction; they're betting on *direction plus timing*.

Sellers take the opposite side of that bet. Time is not a guarantee—markets can move fast and violently—but time is a tailwind. Many short option positions become easier to manage if the underlying behaves even moderately.

This is why premium selling is attractive for income: it can turn a portfolio from "I need the market to reward me" into "I get paid for providing a service, and I manage the risks like a business owner." That emotional shift is bigger than the math.

Actionable takeaway: translate every trade into a one-sentence promise

Before placing any option trade, force yourself to write (or say) this sentence:

> "I collect $___ today, and in exchange I agree that until ___ date, I may be required to *buy/sell* ___ shares at $___ per share."

If the sentence makes you uncomfortable, the trade is too big, too aggressive, or simply not aligned with what you actually want to own. If the sentence sounds like a deal you'd happily sign on paper, you're speaking the language that income options are written in.

4.2 Pricing Drivers and the Greeks

If options were priced by vibes, selling premium would be a religious experience: you'd collect money, whisper a prayer, and hope the market stays polite. Instead, options are priced like insurance and airline tickets—coldly, continuously, and with the unblinking logic of supply and demand. The good news is that you don't need calculus to understand why an option costs what it costs. You just need to know what the market is charging for: *time, uncertainty*, and the possibility of a nasty surprise.

A premium seller makes a living by noticing when those things are expensive—and by never forgetting that "expensive" usually means "there's a reason people are scared."

Option price: two ingredients that behave very differently

An option's market price (the premium) is usually described as:

$$\textit{Option Price} = \textit{Intrinsic Value} + \textit{Extrinsic Value}.$$

Intrinsic value is the "already true" part: if the option were exercised *right now*, how much economic value would that create? Extrinsic value is everything else: time remaining and the market's expectation of how wild the underlying might get before the clock runs out.

Income-oriented option selling mostly harvests extrinsic value. You're not trying to buy a dollar for ninety cents; you're getting paid for carrying risk through time. That means your real product isn't direction. Your real product is *exposure*: exposure to time passing, and exposure to volatility changing.

The pricing drivers: what moves the premium when you're not looking

Option premiums don't sit still. Even if the stock price doesn't move, the option price can. That's not a bug—it's the whole point. The market is constantly updating two questions:

- How much time is left for something to happen?
- How likely is something dramatic to happen?

Practically, an option's price is shaped by a handful of variables:

- The underlying price (where the stock is right now)
- The strike price (the deal price you pre-agreed to)
- Time to expiration (how long the contract is alive)
- Implied volatility (how turbulent the market expects the ride to be)
- Interest rates and dividends (usually second-order for intuition, but not irrelevant)

The Greeks are simply a set of *sensitivities*—dials that tell you, "If this input changes a little, about how much should the option price change?" They are not fortune-tellers. They are a dashboard.

And like any dashboard, they're most useful when you understand the story behind the gauges: theta is the clock, vega is the fear meter, delta is the "how stock-like is this option today?" gauge, and gamma is the "how fast can delta change when the stock moves?" warning light.

Theta: the rent you collect for letting time pass

Theta (Θ) is the option's sensitivity to time—often quoted as the approximate change in the option's price for one day passing, holding other factors constant. Translation: theta is the daily rent.

If you sell premium, theta is the quiet engine that can pay you even on boring days. You can be "right" without being prophetic. You can be right because the calendar did what the calendar always does.

But theta has a personality. It doesn't decay in a straight line. It's not a steady drip like a leaky faucet. It behaves more like the last week before a big deadline: nothing feels urgent... until suddenly everything does.

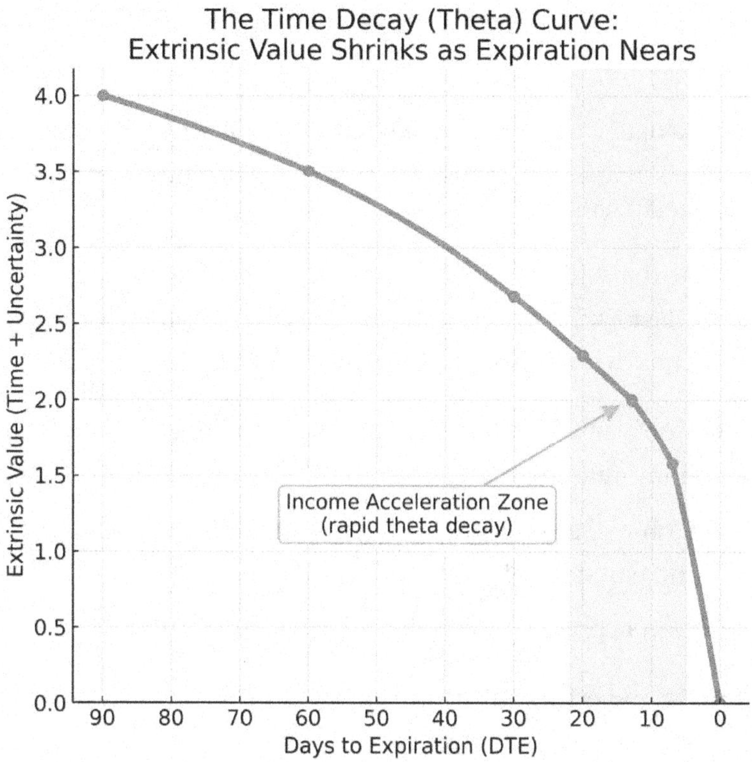

That curve is the reason many premium sellers talk about options the way landlords talk about leases. A 90-day lease doesn't "wear out" evenly. The final week is when keys get returned, deposits get negotiated, and the odds of a dramatic change collapse. Options behave similarly: far from expiration, there's a lot of "maybe." Near expiration, the market has less imagination because there's less time to be surprised.

This is also why selling options can feel psychologically satisfying. The passage of time is an input you can count on. Markets can be irrational, but the calendar is boringly consistent.

The catch with theta: the market doesn't hold other variables constant

Theta is usually quoted "all else equal." Real life is not "all else equal." Volatility changes. Prices gap. Headlines drop at 4:07 p.m. The world is not a controlled experiment.

So treat theta like a paycheck that arrives on schedule *unless* there's a storm. If you sell premium, you're in the storm business. The correct response isn't fear; it's planning:

- Don't sell so close to expiration that one ugly day can dominate the trade.
- Don't size positions so large that you're forced to act at the worst possible time.
- Don't confuse "theta decay" with "guaranteed profit."

Theta is powerful, but it's not magical. It's the rent. You still have to maintain the property.

Delta: not a math symbol---an ``odds" meter

Delta (Δ) is often introduced as "the option's sensitivity to a $1 move in the stock." That's true, but for premium sellers a more useful interpretation is: delta is a rough proxy for the probability that the option finishes in-the-money (with all the caveats that come with models and assumptions).

When you sell an option, you're implicitly betting on *not* being assigned, or at least not being assigned at an inconvenient price. Delta gives you a quick way to talk about that bet without pretending you're omniscient.

A low-delta short option is like writing insurance with a higher deductible: less premium, but more breathing room. A higher-delta short option is like writing insurance close to the cliff: more premium, but less margin for error.

That's why experienced sellers often talk about "selling the 20-delta put" or "the 30-delta call." It's a shorthand for risk posture. Not because delta is perfect, but because it's a common language that forces you to admit what you're doing: choosing how close to the fire you want to stand.

Gamma: the speed at which risk can change

Gamma (Γ) measures how delta changes as the underlying moves. If delta is your speed, gamma is your acceleration.

Gamma is what makes near-the-money, near-expiration options feel like they "wake up" suddenly. A small underlying move can swing delta quickly, which means your position can go from "chill" to "intense" in a hurry.

For an income seller, gamma is the reason the last couple of weeks before expiration can be both attractive *and* dangerous. The extrinsic value may be melting fast (theta tailwind), but the position can also become more sensitive to price moves (gamma risk). It's like driving downhill with a strong wind at your back: you'll get there faster, but your brakes matter more.

If you ever wondered why some traders avoid selling options with only a few days left, gamma is the answer. It's not moral superiority. It's risk management.

Vega: the fear premium you're really selling

Vega (ν or V) measures how much an option's price changes when implied volatility changes by 1 percentage point, all else equal. Translation: vega is your exposure to the market's mood.

Implied volatility is not the same thing as historical volatility. It's not "what the stock did." It's "what the options market is pricing in now"—the collective guess about how turbulent things might get before expiration.

When implied volatility rises, option premiums get more expensive. That's wonderful if you're buying options (you already own an asset that just got pricier? No—you get hurt). Actually, it's wonderful if you *already sold* options and want to close them cheaper? No—rising volatility usually makes it *more expensive* to buy them back. So for the seller, volatility expansion can be the moment when a perfectly calm-looking trade suddenly shows a loss even if the stock hasn't moved much.

That's the signature experience of premium selling: the P&L can temporarily look wrong because fear got repriced.

Here's the deeper truth, though: high implied volatility is often a premium seller's best friend *if* you have the stomach and the discipline to treat it like a sale at an insurance office. When the town is calm, nobody buys flood insurance and premiums are cheap. When the river rises, everyone wants a policy and premiums spike. The seller who only writes policies when nobody wants them isn't a hero—they're just underpaid.

The trick is not "sell volatility because it's high." The trick is "sell volatility when it's high *and* you can afford to be wrong without being forced to liquidate."

Volatility crush vs. volatility expansion: why your timing matters

Two phrases explain a lot of option weirdness:

Volatility crush.
After a big event (earnings, court rulings, FDA decisions, macro announcements), implied volatility often drops because uncertainty collapses. The scary "maybe" becomes a known "did." That drop can make option premiums fall quickly, helping premium sellers.

But volatility crush can tempt you into selling options "right before" the event because premiums look juicy. That's like opening an umbrella rental stand on the beach during hurricane warnings. You'll make great money—unless the hurricane actually arrives.

Volatility expansion.
This is the opposite: implied volatility rises, often during sell-offs or sudden uncertainty. That can inflate option prices and make a short option position show losses quickly. Sometimes it's temporary panic; sometimes it's the market correctly sniffing out real trouble.

Premium sellers live with volatility expansion the way insurers live with storms: you plan for it, you expect it occasionally, and you make sure one storm doesn't end the company.

A seller's dashboard: how the Greeks work together

It helps to think of a short option trade as a bundle of exposures you've chosen:

- *Short theta*: you benefit as time passes.
- *Short vega*: you're hurt if implied volatility rises.
- *Delta exposure*: you have some directional sensitivity, which can change.
- *Gamma risk*: that sensitivity can change faster near the strike and near expiration.

This is why premium selling is not just "collect decay." It's "run a small book of risks." The business can be wonderfully consistent when managed well, and surprisingly violent when managed casually.

A concrete example: why a ``nothing happened" week can still move your P&L

Imagine you sold a put on a steady ETF. The price barely moves all week—your chart looks like a lazy river. You expect your position to slowly profit from theta. Then on Thursday, the market gets nervous about a macro headline. The ETF is still basically flat, but implied volatility jumps.

Your short put can lose value even though your directional call was fine. That's vega: the market is charging more for insurance today than it did yesterday. If you wanted to close the position, you'd have to buy that insurance back at a higher price.

This is the moment when beginners say, "Options are rigged." They're not. You sold something whose price depends on the market's appetite for safety. The appetite changed.

The upside is that the same mechanism can work in your favor. If you sold when fear was high and later fear fades, your short option can profit quickly even without a big move in the underlying. That's not luck. That's understanding what you sold.

Rho, dividends, and rates: the background music

Rho (interest-rate sensitivity) exists, and for certain underlyings and longer-dated options it can matter. Dividends matter too, especially because they can influence early exercise behavior and forward pricing dynamics.

For most day-to-day income option decisions, though, theta and vega are the main characters. Rates and dividends are more like background music: you'll notice them if you listen closely, and sometimes the song changes at a key moment, but they're not the part you hum.

Actionable takeaway: use a ``two-question filter'' before you sell any premium

Before placing a short option trade, answer two questions in plain English:

1. *What am I getting paid for time?* Look at the premium relative to the days remaining and ask, "Is this rent worth the obligation?"

2. *What happens if fear gets more expensive?* Assume implied volatility rises after you sell. If that scenario would force you to panic-close, the position is too large, too close to the strike, or too close to expiration.

If those answers feel sturdy—paid enough for the lease, resilient enough for a mood swing—you're not just selling options. You're pricing risk like a professional.

4.3 Understanding Assignment and Exercise

The word *assignment* has a special talent for making grown adults act like they just found a spider in the shower. It sounds like punishment. Like you did something wrong. Like the broker is about to ring a bell and point at you. In reality, assignment is the most boring kind of drama: it's the contract doing exactly what it said it could do.

The premium was never a gift. It was a receipt. Assignment is what happens when the customer actually uses the coupon you sold them.

Assignment is not a margin call, not a lawsuit, not a surprise bill

Here's the clean translation:

- If you are *short a put* and you get assigned, you *buy* 100 shares per contract at the strike price.
- If you are *short a call* and you get assigned, you *sell* (*deliver*) 100 shares per contract at the strike price.

That's it. No hidden committee. No "gotcha" clause where the price changes. No negotiation. The whole point of listed options is that they're standardized. When assignment happens, the trade becomes stock.

What *can* feel surprising isn't the rule; it's the timing. Many new sellers assume assignment only happens at expiration, like a door that only opens at one specific minute. American-style equity options can be exercised early, which means assignment can arrive on a random Tuesday, usually when you least want another email.

That's not the market being mean. That's the buyer deciding, "I want the stock position now," and the clearing system picking a short seller to fulfill the obligation.

Exercise vs. assignment: two sides of the same coin

People often say "I got exercised," which is like saying "I got shopped." One side chooses; the other side gets chosen.

- *Exercise* is what the *option holder* does. They choose to turn the option into stock at the strike.
- *Assignment* is what happens to the *option writer* (*seller*). They are selected to fulfill the contract.

If you sold the option, you do not get to vote. You can manage the position before it happens, but if a holder exercises and the system assigns you, the contract converts into stock.

That's why premium selling is closer to running an insurance desk than playing a guessing game. When you sell protection, you're agreeing that a claim might arrive. Your job isn't to hope claims never happen; your job is to price policies so claims are survivable.

American vs. European: one key difference that matters in real life

This is one of the few "textbook" distinctions that actually shows up in retail accounts.

- *American-style* options can generally be exercised any time up to expiration. Many equity and ETF options in the U.S. are American-style.
- *European-style* options can generally be exercised only at expiration. Some index options are European-style.

If you're selling American-style options, you're carrying early-assignment risk the entire time the contract is open. In most day-to-day situations, early exercise is uncommon because it often doesn't make financial sense for the holder. But "uncommon" is not the same thing as "never," and the situations where it *does* make sense tend to cluster around one very specific trigger: dividends.

The Friday myth: ``expiration is when everything happens"

Expiration has a certain Hollywood vibe. People imagine a dramatic closing bell where all contracts either disappear or explode. The reality is more procedural, more paperwork, and—if you're prepared—more manageable.

Two things can happen as expiration approaches:

- You close the position yourself by buying back the option (or selling to close if you were long).
- You let it ride, and whatever is in-the-money (ITM) at expiration is likely to be exercised through the industry's automatic procedures, unless the holder gives instructions to do otherwise.

A good premium seller doesn't treat expiration like a cliff. It's a deadline, yes, but it's not the only decision point. You can often choose to exit earlier if the trade has already paid you most of what it's going to pay you.

There's also a practical reason to avoid the "cliff" mindset: stock prices can move after-hours. That creates the weirdest emotional experience in options: thinking you're safe at 3:58 p.m., then seeing the underlying jump at 4:12 p.m., and wondering if your option that looked dead just came back to life.

What actually happens when you're assigned

Assignment is operationally simple, even if it feels emotionally loud.

If you're assigned on a short put.

Your account buys 100 shares per contract at the strike price. Your cash decreases by 100 × *strike* per contract (net of the premium you already received), and you now own shares with a cost basis tied to the strike and the premium.

That last part matters: the premium you collected effectively reduces your net purchase price. If you sold a $50 put for $1.20 and get assigned, you're buying at $50, but your net cost is

$48.80 per share in the simplest economic sense (ignoring commissions, taxes, and other specifics).

If you're assigned on a short call.

Your account delivers 100 shares per contract at the strike price. If you already own the shares, they're sold away at that strike (covered call assignment). If you do not own the shares, you become short stock (a very different risk profile that brokers may or may not allow depending on account permissions).

This is where the "obligation" word earns its keep. The contract doesn't care what you *meant*. If you sold a call without owning shares and get assigned, the system still expects delivery. The mechanical outcome can be a short stock position that changes your margin requirements overnight.

Income strategies often use covered calls precisely to keep this outcome sane: if you're going to have an obligation to sell shares, it's comforting to already have the shares.

Assignment is random (and that's a feature, not a flaw)

One of the most unsettling questions beginners ask is, "Why me?" The answer is: you weren't singled out. Assignment is generally processed through a clearing and allocation system designed to be fair and operationally efficient. You sold a contract in a standardized pool, and when someone exercised that series, a short position was selected according to the broker's and clearing firm's procedures.

This randomness is why "I'll just avoid assignment by being careful" is not a plan. The plan is: assume assignment can happen whenever it is rational for a holder to exercise, then structure your trades so assignment is an acceptable outcome.

That single mental move—treating assignment as a normal transaction instead of a failure—changes the entire stress level of premium selling.

Early exercise: why it's usually rare, and why dividends change the math

Most early exercise is irrational for calls, because the call holder would be giving up remaining extrinsic value by exercising early. If the option still has time value, exercising is like tearing up a gift card and insisting on cash today, even though the gift card could be sold for more than the cash value.

But dividends create a loophole in that logic.

Dividend risk for short calls.

If you are short a call on a dividend-paying stock, you face a specific risk window around the *ex-dividend date* (the date when buyers no longer receive the upcoming dividend). A call holder might choose to exercise early to own the stock before the ex-dividend date and collect the dividend.

Whether that makes sense depends on a comparison:

- the dividend the holder can capture by owning the shares, versus
- the extrinsic value they would forfeit by exercising early.

If the dividend is larger than the remaining extrinsic value (and financing costs, depending on the situation), early exercise becomes more likely. That's when covered call sellers can wake up to find shares called away earlier than expected.

This is not "bad luck." It's the buyer acting rationally. You sold someone the right to buy; occasionally they buy early because a dividend makes it attractive.

The practical lesson is almost disappointingly simple: if you sold calls on a dividend stock and your short call is in-the-money with very little extrinsic value left, pay attention as the ex-dividend date approaches. That's the high-risk zone for early assignment.

Early exercise for puts.

Early exercise for puts can happen too, particularly when rates, borrow considerations, or deep-in-the-money conditions make it attractive. It's less commonly the "gotcha" for retail income strategies than dividend-driven call exercise, but it exists. The main point is the same: early exercise is most likely when the option has little time value left and a stock position is economically preferable to holding the option.

Expiration mechanics: the quiet power of automatic exercise

Most people assume an option expires and disappears unless someone does something heroic. In practice, there are industry procedures that may automatically exercise certain in-the-money options at expiration unless the holder instructs otherwise.

That's a big deal for sellers because it means: if your short option finishes in-the-money, you should assume it can turn into stock without anyone calling you for permission.

The industry also supports the idea of "contrary instructions"—the holder can generally communicate "do not exercise" or "exercise anyway" depending on the situation. Brokers often have cut-off times for receiving those instructions, and those deadlines can be earlier than the official exchange cut-offs. Translation: the clock you care about is usually your broker's clock.

This is where the "Friday afternoon drama" actually comes from. It's not the math; it's the operational schedule:

- Prices can move quickly late in the day.
- Customers have decision deadlines.
- Brokers have processing deadlines.
- The clearing system has its own cut-offs.

When you understand that expiration is a process—not a single instant—you start behaving like a professional: you decide ahead of time what you'll do if the option is hovering near the strike, and you don't leave important outcomes to the final minutes.

``Pin risk'': the uncomfortable zone near the strike

There's a special kind of annoyance that can happen when the stock closes very close to your strike at expiration. The option may appear to be out-of-the-money by a penny at the close, then drift in after-hours. Or it may appear in-the-money and then drift out. Meanwhile, exercise decisions and automatic thresholds are being applied through a system you don't fully control.

This is why many income traders don't like to "wing it" when an option is near the strike late on expiration day. It's not superstition. It's an acknowledgment that tiny price differences can create a very non-tiny change in what you own on Monday morning.

If assignment would be perfectly fine-because you wanted the shares anyway, or you're happy to sell them at the strike anyway-pin risk is more of a nuisance than a threat. If assignment would create a margin scramble or an unwanted short stock position, pin risk is a flashing sign that your position management needs tightening.

Multi-leg strategies: assignment doesn't fix your trade for you

Spreads and combinations can create a false sense of safety: "If I'm assigned on one leg, the other leg will protect me automatically." It won't. Options are contracts, and contracts don't coordinate out of kindness.

If you're short one leg of a multi-leg strategy and it gets assigned, the remaining leg stays exactly what it is until you take action. That can produce temporary stock positions and margin effects you didn't plan for.

This isn't a reason to fear spreads; it's a reason to respect plumbing. If you trade multi-leg structures for income, you want to know what you will do if one leg turns into stock—especially near expiration and around dividends—because the market won't pause to ask if you intended to be elegant.

Making assignment boring: the income seller's ideal outcome

The goal isn't to "never get assigned." The goal is to make assignment emotionally neutral.

For short puts:

Assignment should feel like, "Fine. I bought the shares at the price I was paid to agree to." If you sold the put as a way to enter a stock you'd happily own, assignment is just the entry method being completed.

For covered calls:

Assignment should feel like, "Fine. I sold at my pre-chosen price and kept the premium." If you sold the call because you were willing to sell the shares at that level, assignment is simply the sale happening on schedule—or sometimes a bit early because dividends made it rational for the holder.

When assignment feels like an ambush, it usually means the strike was chosen for premium rather than for sanity. Premium is seductive; sanity is profitable.

Actionable takeaway: adopt an ``assignment-ready" checklist before you open the trade

Before selling any option, answer these questions without flinching:

- *If I'm assigned tomorrow, am I okay with the stock position I'll have?* (Long shares for short puts; short shares for uncovered short calls; sold shares for covered calls.)

- *Do I understand the one scenario where early exercise is most likely?* For covered calls on dividend stocks, assume ex-dividend dates matter when your call is in-the-money and low on extrinsic value.
- *If the option is near the strike late on expiration day, will I manage it proactively rather than "see what happens"?* Decide ahead of time whether you'll close, roll, or accept assignment.

If those answers are clean, assignment stops being a horror story and becomes what it really is: the paperwork of getting paid to make a deal.

4.4 Collateral and Buying Power

The most expensive sentence in options trading is also the most common: *"I had the trade right...but I got margin called."* That's not tragedy. That's plumbing. Your thesis can be brilliant and your account can still be forced to do something humiliating at the worst possible time—because you ran out of collateral.

Premium selling rewards patience, but it punishes fragility. If your positions depend on your broker continuing to "feel comfortable," you don't really own a strategy. You own a fair-weather agreement.

Collateral: the security deposit you don't get to argue about

Collateral is the good-faith deposit your broker requires to let you open a position. With short options, you're selling a contract that can turn into stock obligations quickly. The broker is on the hook operationally if you can't perform, so the broker demands a buffer.

That buffer is not a moral judgment. It's not your broker "being mean." It's the broker protecting itself from the very real possibility that your obligation becomes large at the very moment your account is shrinking.

A helpful way to think about collateral is to imagine you're renting out a car. You can be the world's nicest rental company, but you still swipe a credit card for a deposit. Not because you assume the customer will crash the car—because the *deposit* is what makes the agreement enforceable when something goes wrong.

Buying power: your account's shock absorbers

Buying power is what your broker says you can still deploy for new positions without violating its risk requirements. It's easy to treat buying power like "cash I haven't used yet." It isn't. Buying power is *permission*, and permission can be revoked or reduced when markets get volatile.

This is where premium selling can feel unfair to beginners. You collect cash when you sell an option—real, spendable cash—but your broker may simultaneously lock up collateral that is larger than the premium you received. The premium is your income; the collateral is your responsibility.

If you only look at the cash inflow, you'll be tempted to scale too quickly. If you look at the buying power impact, you'll understand what the broker thinks you just promised.

Cash account vs. margin account: two different universes

Two investors can sell the same option and experience completely different "costs" in their account depending on account type.

Cash account. A cash account is the strict parent. If you sell a put in a cash account, the broker typically expects you to have enough cash to buy the shares at the strike price. That's why people call it a *cash-secured put*. You're literally securing the obligation with cash.

This can feel boring, which is exactly why it's such a strong foundation for income strategies. Boring is resilient.

Margin account. A margin account is the lenient parent with a long list of house rules. Instead of requiring the full cash amount for a short put, the broker may require a formula-based margin amount. That requirement changes with the underlying price, the option's moneyness, and market volatility.

This can be useful. It can also be addictive. The danger isn't "margin exists." The danger is thinking margin is stable. Margin is stable until it isn't.

Cash-secured puts: getting paid to place a limit order (with receipts)

A cash-secured put is best understood as a paid limit order.

Imagine you want to buy 100 shares of a stock at $50, but it's currently $53. You could place a limit order at $50 and wait. A cash-secured put is a different version of that same intention:

- You sell the $50 put.

- You collect premium today.
- If the stock drops and you're assigned, you buy at $50.

The difference is psychological and financial. A limit order pays you nothing for waiting. The put sale pays you for waiting. That premium is compensation for agreeing to buy during discomfort—because if you're assigned, it likely means the stock fell and the mood is bad.

Cash-secured puts have a beautiful discipline baked in: your position size is constrained by actual cash. If you can't afford 100 shares at the strike price, you can't pretend you can. That alone prevents a large percentage of catastrophic beginner mistakes.

Covered calls: the other ``cash-secured" trade

A covered call is collateralized by shares you already own. When you sell a call against 100 shares, you are telling the broker, "If I'm assigned, I will deliver these shares."

That's why covered calls are often treated more gently from a margin perspective than naked short calls. The broker can see the shares. The broker can deliver them. The obligation is funded.

This is also why covered calls can be emotionally calming for income investors. Your worst-case assignment outcome is often a sale at a price you chose. That might be disappointing if the stock rockets higher, but it isn't existential.

Uncovered (naked) short options: the words matter

A short option that is not backed by cash (for puts) or shares (for calls) is often called *uncovered* or *naked*. The nickname is accurate: your obligation is exposed.

Uncovered short calls are particularly dangerous because the underlying can, in theory, rise without limit. If you are obligated to sell stock at a strike price while the market price keeps climbing, your losses can grow fast and ugly. Brokers know this, which is why requirements can be steep and why forced liquidations can happen quickly.

Uncovered short puts can also be dangerous, not because losses are infinite, but because the obligation can still be enormous relative to your account size. A stock doesn't need to go to zero to create a life-changing drawdown if you oversold contracts.

Premium selling is a business of small edges. Uncovered short options are where people try to turn small edges into big thrills. That's usually where the account learns the difference between "I got paid" and "I was adequately capitalized."

Why your premium shows up, but your money isn't ``freed"

One of the most confusing first experiences in option selling is this:

> "I sold a put and collected $200. Why did my available buying power drop by $3,000?"

Because the broker isn't running a charity; it's running a risk engine. The premium is not a down payment that replaces collateral. It's income, yes, but it doesn't erase the fact that you might have to buy shares. Depending on the account type and the broker's rules, the premium may help satisfy part of the requirement, but it won't eliminate it.

If this feels restrictive, that's the point. Collateral requirements are the guardrails that keep premium selling from becoming a "sell everything" game.

Buying power expansion: the trap door during volatility

Here's the part that tends to surprise even experienced investors: collateral requirements can *expand* in volatile markets.

When markets get ugly, two things often happen at once:

- Your positions move against you (stock down for short puts; stock up for short calls; volatility up for many short options).
- The broker demands more collateral because the position is now riskier.

That second piece is the trap door. You can be losing money and losing buying power simultaneously. It's like driving a car downhill and having the road get narrower at the same time.

This is why "I'll just roll it" is not a universal solution. Rolling often requires buying back the current option and selling a new one. If your buying power is constrained, you may not be able to execute the roll cleanly or at attractive prices. The market doesn't care that rolling is your preferred coping mechanism.

The practical rule is simple: if your strategy requires you to always have abundant buying power, your strategy is more fragile than it looks.

House margin: your broker can change the rules mid-game

Regulatory frameworks provide baselines for margin requirements, but brokers can—and often do—impose additional "house" requirements. They may do it for certain volatile tickers, for concentrated positions, around earnings, or during broad market stress.

That's not a conspiracy. It's a reminder that your broker is not your partner in the trade. Your broker is the risk manager for the loan it extends and the obligations it helps you clear.

The take-home lesson is not to memorize every formula. The lesson is to treat margin as *variable* and size positions so you can survive an unpleasant change in requirements.

The hidden danger: concentration feels efficient until it becomes fatal

Selling options can create a false sense of diversification. You might have ten short puts on ten different tickers and think, "I'm spread out." Then a market-wide volatility spike hits and every one of those positions demands more collateral at once. Diversified pain is still pain, and collateral calls don't care that you have ten tickers instead of one.

Concentration shows up in sneaky forms:

- Too many contracts in the same sector (all your "different" stocks are actually one macro bet).
- Too many positions with the same expiration (all your risk clusters in the same week).
- Too much reliance on a single market regime (calm markets) to keep requirements low.

Income strategies work best when they are built to be *survivable* across regimes—because markets have no obligation to stay calm just because your spreadsheet likes calm.

A practical sizing mindset: treat collateral like oxygen

A healthy income account is not maximally invested. It is comfortably invested. It has slack.

Slack is what allows you to:

- hold through volatility without panic-closing,
- roll or adjust without begging for buying power,
- take assignment when it's part of the plan,
- avoid forced selling when prices are worst.

If that sounds conservative, good. Premium selling is not a sport where the bold get medals. It's closer to running a small factory: steady inputs, steady outputs, and a deep suspicion of any week that looks "too profitable."

Cash-secured vs. margin: a clean comparison that isn't moralistic

It's tempting to treat cash-secured puts as "good" and margin as "bad." The real distinction is *stability of obligations*.

Cash-secured structures tend to have:

- clear maximum commitments,
- predictable funding needs,
- fewer forced decisions under stress.

Margin-based approaches can offer:

- capital efficiency,
- flexibility across multiple positions,
- the ability to run diversified strategies with less idle cash.

The tradeoff is that margin adds a third party's risk tolerance to your process. Not just the market's volatility—your broker's response to it. That's a real risk factor, and it belongs in the same mental bucket as vega: it can expand when fear expands.

When ``income'' becomes leverage: the premium-reinvestment temptation

One of the most seductive loops in options selling goes like this:

> "I sold options, collected premium, and now I have more cash. I can sell more options."

This is how people accidentally turn an income strategy into a leverage strategy. The premium feels like a renewable resource. In calm markets, it *acts* like one. Then a volatility spike arrives, premiums widen, losses appear, requirements rise, and suddenly last month's "extra cash" is not extra at all—it's the buffer you needed to keep control.

There's nothing wrong with reinvesting income. The mistake is reinvesting it in a way that increases your contractual obligations faster than your account can absorb stress.

A good rule of thumb: treat premium like a *wage*, not a *loan*. Wages improve your life slowly. Loans change your life quickly—for better or worse.

Actionable takeaway: adopt a ``survival-first" collateral rule you can follow blindly

Before you open any new short option position, run this simple check:

1. *Assume collateral requirements increase sharply and immediately.* If that would push you near a margin call, the position is too large.
2. *Assume you get assigned on your short puts.* If you can't comfortably own the shares (cash-secured) or finance them without stress (margin), the strike is fantasy.
3. *Keep deliberate slack.* Decide on a minimum percentage of buying power you will not use—your "oxygen reserve"—and don't negotiate with yourself when premiums look tempting.

That's how premium selling stays an income factory instead of becoming a one-time fireworks show: you never let the business run out of oxygen.

4.5 Reading the Chain and Selecting Underlyings

An option chain is a menu where the prices are real, the portions are huge, and the font is designed by someone who dislikes human happiness. The numbers move. The columns multiply. One wrong click and you've accidentally sold insurance on a stock you've never heard of because it had a "nice premium." That's how people learn an expensive lesson: the chain doesn't care whether you *meant* to be careful.

The goal here isn't to become fast. The goal is to become *unfooled*. A good income trade is rarely the one with the biggest premium. It's the one you can enter and exit without donating half the paycheck to the bid/ask spread, and the one that won't turn your account into a hostage situation when volatility spikes.

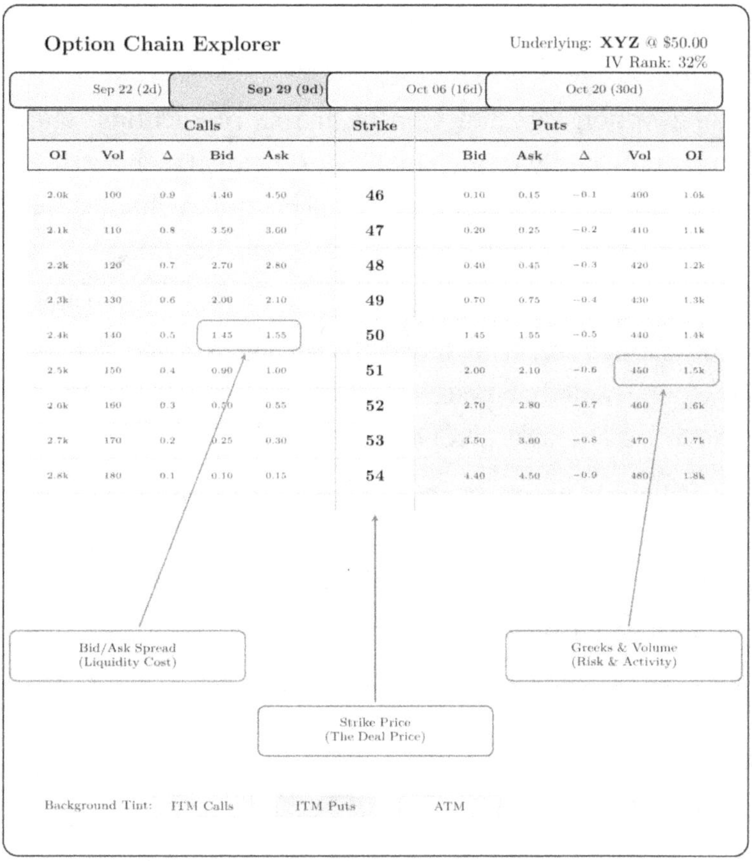

Calls					Strike	Puts				
OI	Vol	Δ	Bid	Ask		Bid	Ask	Δ	Vol	OI
2.0k	100	0.9	4.40	4.50	46	0.10	0.15	−0.1	400	1.0k
2.1k	110	0.8	3.50	3.60	47	0.20	0.25	−0.2	410	1.1k
2.2k	120	0.7	2.70	2.80	48	0.40	0.45	−0.3	420	1.2k
2.3k	130	0.6	2.00	2.10	49	0.70	0.75	−0.4	430	1.3k
2.4k	140	0.5	1.45	1.55	50	1.45	1.55	−0.5	440	1.4k
2.5k	150	0.4	0.90	1.00	51	2.00	2.10	−0.6	450	1.5k
2.6k	160	0.3	0.50	0.55	52	2.70	2.80	−0.7	460	1.6k
2.7k	170	0.2	0.25	0.30	53	3.50	3.60	−0.8	470	1.7k
2.8k	180	0.1	0.10	0.15	54	4.40	4.50	−0.9	480	1.8k

The chain is doing two jobs at once. It's *quoting prices* (bid/ask) and it's *broadcasting a mood* (implied volatility, deltas, and the rest of the model outputs). Most bad option trades start when someone reads it like a lottery ticket dispenser: "Which one pays the most?" Most good trades start when someone reads it like a business owner: "Can I sell this policy at a fair price, in a market where I can exit without getting mugged?"

The only reason to care about the chain: execution and survivability

Income options are not won by heroic predictions. They're won by repeatedly doing small, sensible deals where the friction is low and the exit is realistic. The option chain is where you spot friction before it quietly eats your returns.

If you've ever made a trade and immediately felt, "Why am I down already?", you've met the bid/ask spread. The chain shows you whether you're about to pay a toll-sometimes a small toll, sometimes a robbery.

It also shows you whether you're playing in a crowded, liquid market (good) or an empty, thin market (dangerous). Thin markets don't just cost more to enter. They can trap you when you need to leave.

Start with the underlying: don't sell premium on a stock you wouldn't date

A premium seller is not just selling a contract; they're volunteering to become a shareholder (puts) or to sell a position at a fixed price (covered calls). That means your underlying selection is not a side detail. It's the foundation.

A simple rule keeps you out of most disasters: *don't sell puts on something you'd be ashamed to own if the market closed for a year*. If the company's story makes you nervous, the option premium is not "income." It's a bribe.

For covered calls, a similar rule applies: don't sell calls on a stock you can't stand to see sold away at your strike. If losing the shares would feel like losing a family heirloom, don't write a contract promising to hand them over.

Good underlyings for income option selling tend to have:

- deep, consistent trading volume in the shares,
- active options markets across multiple strikes and expirations,
- relatively stable corporate behavior (no constant "surprise!" events),
- spreads that don't widen into a horror movie the moment volatility rises.

None of that guarantees profit, but it does something more important: it makes your strategy *operationally feasible*.

Reading the chain like a pro: bid, ask, and the invisible tax

Every option has two prices:

- *Bid*: what someone will pay you right now.
- *Ask*: what someone will sell to you for right now.

The spread is the difference. And that spread is a hidden tax you pay when you enter *and* when you exit. If you sell an option at the bid and later buy it back at the ask, the spread is the first chunk of premium you effectively give away.

For income strategies, spread matters because you're not hunting one giant win. You're stacking many modest wins. A wide spread is a leak in the factory roof.

A practical mental trick: before selling any contract, pretend you'll need to buy it back immediately. If that thought makes you wince because the ask is dramatically higher than the bid, you're not looking at "income." You're looking at a market with poor competition.

Volume and open interest: ``Is anyone else playing?"

Two columns in most chains answer the most important liquidity question.

Volume. Volume is how many contracts traded today. It's a snapshot of current activity.

Open interest. Open interest is how many contracts exist and are currently open (not closed out). It's a snapshot of participation over time.

For an income trader, you're not just asking, "Can I get filled?" You're asking, "Can I get out later at a fair price?" Higher open interest and steady volume usually help. They don't guarantee tight spreads, but they often correlate with healthier markets.

A thin option with a wide spread can look seductively profitable because the mid-price is generous. But if you can't realistically transact near the mid, that "profit" is imaginary. It's like valuing your house based on the highest listing in the neighborhood rather than what buyers actually pay.

Implied volatility: the price tag on fear

Implied volatility (IV) is not a moral score. It's the market's price for uncertainty. High IV can be attractive to premium sellers because it means you're paid more for the same strike and expiration. But high IV also means the market believes something more violent is plausible.

Treat IV like you'd treat weather forecasts. A hurricane warning makes umbrellas expensive. Selling umbrellas at hurricane prices can be profitable—if you have a plan for the hurricane. If you don't, you're just the person standing outside in the wind holding inventory.

In practice, IV helps you answer two questions:

- Is premium unusually rich right now, or is it skinny?
- If volatility expands further after I sell, can I live with the mark-to-market pain?

Because you're short vega when you sell premium, IV isn't just a "nice to have" number. It's the dial that tells you how sensitive your trade might be to the market getting more nervous.

Delta in the chain: a useful probability-ish compass (not a promise)

Most chains show delta. For premium sellers, delta is a practical shorthand for "how close am I standing to assignment territory?" A higher delta means the option behaves more like stock and is more likely to finish in-the-money, all else equal.

The chain makes delta feel precise-three decimal places can look like certainty. It isn't certainty. Delta is a model output that assumes a set of conditions that can change. But it's still useful because it helps you compare candidates quickly:

- a 0.15 delta short put is usually "farther out," lower premium, higher probability of expiring worthless,

- a 0.35 delta short put is usually "closer in," higher premium, higher chance of becoming a stock position.

The important part is that delta forces honesty. It turns "I feel safe" into "I chose a strike with X level of exposure."

Expiration selection: the time you sell is part of the trade

Every chain is a grid: strikes across one axis, expirations across the other. Many beginners choose expiration by convenience ("next Friday is fine"). That's like choosing a lease length because the calendar looks tidy.

Expiration choice changes:

- how much theta you have working for you per day,
- how much gamma risk you take on near the strike,
- how much time you give the underlying to wiggle into trouble,
- how much flexibility you have to manage without getting cornered by the deadline.

Short-dated options can feel delicious because the premium "decays fast." They can also be unforgiving: with less time, small moves matter more, and management decisions become urgent. Longer-dated options can be calmer operationally, but you're selling more time and more vega exposure.

A premium seller doesn't need a single "best" expiration. They need consistency: choosing expirations where the trade can breathe, and where the P&L isn't dominated by the last two days of chaos.

The three checks before you click: spread, participation, and honest premium

Most "trap" trades fail one of these checks:

1) **Liquidity via spread.** If the spread is wide relative to the premium, the market is telling you something: there isn't enough competition to give you fair pricing. For income strategies, that's poison. You're trying to collect rent, not pay tolls.

2) **Participation via volume and open interest.** Thin markets can trap you during volatility. When you most need a clean exit, the spread can widen, fills can get ugly, and your "manage it later" plan can become expensive.

3) **Premium quality via IV context.** Premium isn't "good" because it's large. Premium is good when it's large *relative to the risk*. IV helps you sense whether the market is paying you appropriately-or whether you're being bribed to step in front of a moving truck.

These checks don't require sophistication. They require discipline. They're the difference between professional boring and amateur exciting.

Selecting underlyings: the income trader's ``no-fly" list

A practical way to improve results is to build a personal no-fly list-categories of underlyings where option income is usually more trouble than it's worth.

Common suspects:

- *Illiquid small caps* with options that barely trade.

- *Meme-driven names* where price can move 20% on vibes and a tweet.
- *Earnings roulette* when you're not intentionally trading earnings (premiums can be rich for a reason).
- *Wide-spread chains* where "mid" pricing is fantasy and exits are punitive.

Avoiding these isn't cowardice. It's protecting the factory. Your edge is not that you can predict chaos better than everyone else. Your edge is that you can run a process that doesn't break when chaos arrives.

A quick case study: two identical premiums, two totally different trades

Imagine you see two short puts with the same premium-$1.00 for a 30 DTE contract. One is on a high-volume ETF with a $0.02 wide spread and heavy open interest. The other is on a thinly traded stock with a $0.25 spread and sparse open interest.

They look identical if you only stare at the premium. They are not identical in the real world.

The ETF option is like selling insurance in a big city: lots of buyers, lots of sellers, competitive pricing, and a decent chance you can adjust without drama. The thin option is like selling insurance in a town with one road in and one road out: if the weather changes, you're stuck with whatever terms the market offers.

Premium selling is a game of *repeatability*. Anything that makes exits unreliable is an enemy of repeatability.

Mechanics that save money: limit orders and patience

Income traders often bleed money by using market orders in options. Options spreads and liquidity can change quickly, and market orders can get filled at unpleasant prices-especially in fast markets.

A simple habit can improve execution meaningfully: use limit orders, and be willing to not get filled. If you're selling premium, you're offering a product. You get to name a price. You don't need to accept the first offer.

Patience is not just a personality trait here; it's a trading edge.

Actionable takeaway: a 60-second chain routine before every trade

Before you sell any option for income, take one minute and do this routine:

1. *Spread sanity check:* Is the bid/ask tight enough that you're not donating a chunk of premium to friction?
2. *Liquidity check:* Is there real volume/open interest at your strike and expiration, or are you trading in an empty room?
3. *IV reality check:* Is premium rich because the market is nervous? If volatility expands further, can your account and your nerves handle it?
4. *Assignment honesty check:* If this turns into stock, is that acceptable—financially and emotionally?

If the chain passes those tests, you're not just picking an option. You're selecting a market where your income strategy can actually function-where you can enter cleanly, manage calmly, and exit without paying ransom.

Chapter 5

Operating the Production Lines

Your factory is built, but machines don't run themselves. It is time to flip the switch. In this chapter, we stop waiting for quarterly dividends and start manufacturing our own paychecks using three specific 'machines': the Covered Call, the Cash-Secured Put, and the Credit Spread. We will cover exactly how to operate each one, how to dial in the settings for safety, and what to do when the assembly line jams.

5.1 Covered Calls: Renting Out Your Assets

Picture a friend who owns a small duplex in a solid neighborhood. The rent checks aren't glamorous, but they're steady, and they arrive whether or not the property's Zillow estimate is up this month. That's the mental posture a covered call tries to give you with a stock you already own: not a prophecy about where prices are headed, but a way to get *paid for holding inventory*.

A covered call is simple on the surface: you own at least 100 shares of a stock (or ETF), and you sell ("write") a call option against those shares. The option buyer pays you cash up front (the *premium*). In exchange, you take on an obligation: if the buyer exercises the option, you must sell your shares at a predetermined price (the *strike price*) on or before a predetermined date (the *expiration*). The shares you already own are the "cover"-you're not scrambling to find stock if assignment happens, because you already have it sitting on the shelf.

That shelf-and-rent framing matters because it flips the emotional script. Many investors treat their stocks like lottery tickets: every day is a referendum on whether they were smart. Covered calls treat your stocks like productive assets: you can still benefit from the asset appreciating, but you also demand that it pay you *while you wait*. It's not thrilling in the way a breakout rally is thrilling. It's satisfying in the way a well-run business is satisfying.

The deal you're making: cash today, ceiling tomorrow

A call option is the right (for the buyer) to purchase shares at the strike price. When you sell a call, you are selling that right. Your payment is immediate; your obligation is conditional.

That obligation creates the two signature features of the covered call payoff:

- *A cap:* above the strike price, your upside is limited because you may be forced to sell at the strike even if the market price is higher.
- *A buffer:* the premium you collect cushions small declines. If the stock drops a bit, you can still be ahead because you got paid up front.

This is the whole bargain in one sentence: you trade away the possibility of unlimited upside in exchange for immediate income and a small shock absorber on the downside.

An example makes it concrete. Suppose you own 100 shares at $100. You sell a 30-day call with a $105 strike for $2.00 per share. You collect $200 immediately.

If the stock rockets to $120, you don't get the whole ride. At expiration, the call will almost certainly be exercised, and you'll sell at $105. Your stock profit is $5 per share, plus the $2 premium, for $7 total. You made money. You just didn't make *all* the money you would have made holding the stock naked.

If the stock drifts to $103, nothing gets exercised. You keep the shares and the $200. That $200 is real-cash that can be reinvested, withdrawn, or simply counted as produced income.

If the stock dips to $99, you're down $1 on the shares, but up $2 from the premium. You're net up $1. That is the buffer doing its job.

And *if the stock falls hard*, the strategy doesn't perform miracles. If the stock drops to $80, you are still very much long the stock. You're down $20 on the shares and up $2 from the premium. The covered call is not a helmet; it's a seatbelt.

That last point is where many people misunderstand covered calls. The premium can feel like a "win" independent of what the stock does-especially because it arrives first. But economically, you are still an equity owner. The premium reduces the cost basis and softens modest drawdowns; it does not erase equity risk.

Seeing the shape: the cap and the income buffer

The fastest way to trust a strategy is to see its payoff profile-the geometry of how it behaves as the stock moves. Compare two lines:

- *Stock only:* profit/loss moves one-for-one with the stock.
- *Covered call:* profit/loss is shifted upward by the premium (the buffer), but flattens out above the strike (the cap).

The "income buffer" is the small region where the stock price can fall and you're still not losing money overall because the premium is offsetting the decline. It's not magic; it's math. But it *feels* like something powerful in real life because it changes how you experience normal market noise. A $1 dip that used to be annoying becomes a shrug.

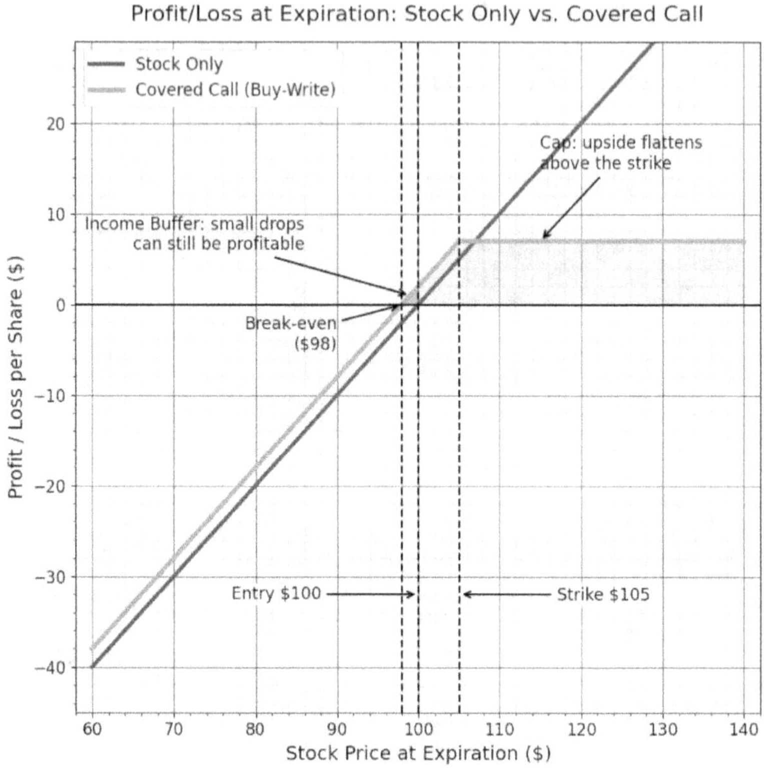

Once you internalize that picture, covered calls stop feeling like a trick and start feeling like a dial you can turn. Want more income today? You typically choose a strike closer to the current price. Want more room for upside? You typically choose a strike farther away. Either way, the cap and buffer are always there-just in different proportions.

Covered calls as cost-basis engineering (not market prediction)

A subtle benefit of the covered call is that it converts some of your future hoped-for returns into current realized cash flow. That sounds like financial alchemy until you notice that many investors already do this in a clumsier way: they sell a little stock to "take some off the table" or to fund a withdrawal. A covered call can be a cleaner mechanism because it doesn't force you to reduce the share count immediately. It asks the market to pay you to keep holding-*and only sells the shares if the market rises enough to justify it*.

This is why covered calls pair so naturally with income-focused portfolios. You're not trying to nail tops and bottoms. You're trying to put your portfolio on salary.

But there's a discipline hidden in this framing: to sell covered calls well, you must be willing to sell the shares at the strike. If you're emotionally attached to the stock-if the idea of it being "taken away" makes you defensive-then you'll either avoid good opportunities or roll yourself into knots later trying to prevent assignment at any cost. The right emotional stance is closer to a shopkeeper than a collector: inventory comes in, inventory goes out, the point is margin and repeatability.

The three outcomes, told like a business ledger

Covered calls are often taught as a branching diagram. In practice, they feel like three different kinds of monthly bookkeeping:

1) The call expires worthless (you keep the shares).

This is the cleanest cycle. You sold a promise, the promise never got called upon, and you kept the payment. The stock might be flat, slightly up, or even slightly down-it doesn't matter. The premium is yours.

Many investors underestimate how psychologically valuable this is. Getting paid for time passing is the opposite of the usual investing experience, where time passing without price movement feels like wasted effort. Here, time is not a thief; time is an employee.

2) The stock rises above the strike and you get assigned (you sell the shares).

Assignment isn't a punishment; it's the contract working. If you chose a strike you're genuinely happy selling at, assignment is an orderly exit with a cash bonus attached.

The hidden win is that assignment *forces* you to take a profit. Humans are famously bad at that. We hold winners too long because selling feels like admitting the party might continue without us. Covered calls can turn that weakness into a rule: "I'll sell at this price, and I'll get paid for making that commitment."

3) The stock falls (you keep the shares, but the premium cushions the drop).

This is where honesty matters. Covered calls do not turn a bad stock into a good investment. They merely change your entry economics and your ongoing cash flow.

If your underlying holding is something you'd be comfortable owning through a drawdown, the premium can make that drawdown easier to endure. If the underlying is a fragile story stock where bad news can permanently reset valuation, the covered call premium is often just a thin layer of paint on a cracked wall.

Picking the underlying: you're choosing a tenant as much as a rent check

Options premium is not free money; it's the market's price for uncertainty. The stocks with the fattest premiums are often the ones most likely to hurt you. That doesn't mean you must avoid them entirely, but it does mean you should treat premium as a signal. If a call is paying you a surprising amount, ask what kind of storm clouds are in the forecast: earnings, product announcements, lawsuits, regulatory rulings, takeover rumors, sector blowups.

A practical way to choose covered call candidates is to start with the stock selection standards you'd use even if options didn't exist:

- businesses you understand well enough to hold without daily reassurance,
- balance sheets and competitive positions that don't rely on perfect conditions,
- liquidity (tight bid/ask spreads on options; active chains).

Then add one more standard that's unique to covered calls: you must be willing to sell the position in 100-share increments. Covered calls are not made for "I never want to sell this" holdings. If you truly never want to sell, then your ceiling is infinitely high—and you shouldn't sell it away.

Choosing the strike: how much ceiling are you willing to give up?

Strike selection is where the "renting out" metaphor becomes concrete. A landlord can charge high rent and risk higher turnover, or charge a little less and keep a good tenant for years. With covered calls, a closer strike generally pays more premium but increases the probability that you'll be assigned (your shares get called away). A farther strike generally pays less premium but gives you more upside room.

Instead of thinking "bullish" or "bearish," think in terms of *acceptable outcomes*:

- If you would be delighted to sell at a specific price, that price is a candidate strike.
- If selling there would make you feel regretful and impulsive, it's not a good strike (or it's not a good stock for you to be writing calls on).

Many investors accidentally choose strikes by staring at the premium and ignoring the ceiling. The premium is visible and immediate; the ceiling is hypothetical and emotional. But the ceiling is the true cost. If you don't price it consciously, you'll pay it unconsciously later.

A simple sanity check helps: write down the *effective sale price* if you get assigned. That's the strike plus the premium collected.

If you sell the $105 call for $2, your effective sale price is $107. Ask yourself: "Would I place a limit order to sell my shares at $107 within the next month?" If the answer is no, you're not writing income-you're writing a future argument with yourself.

Choosing the expiration: selling time without selling your life

Expiration is the other dial. Shorter expirations can feel like faster paychecks-more frequent premium collection, more opportunities to adjust. Longer expirations can offer bigger single credits and fewer decisions.

But there's a human factor most strategy write-ups ignore: decision fatigue. Covered calls are not hard, but they are *recurring*. You're running a process. If the process demands attention every other day, you'll either burn out or start making sloppy decisions just to stop thinking.

Many income-focused traders gravitate toward a rhythm where the option has enough time value to pay decently, but not so much that you're locking in a ceiling for ages. The point is not to be constantly "doing trades." The point is to build a repeatable production line that you can operate while still having a life.

The mechanics that surprise people: assignment can happen early

A covered call seller often assumes assignment happens only at expiration. In reality, assignment can happen any time the option holder chooses to exercise (for American-style equity options). Most of the time, early exercise isn't economically rational because the call holder would be giving up remaining time value. But "most of the time" is not "always," and the exceptions cluster in predictable places.

The most common practical scenario is around dividends. If you're short an in-the-money call and a dividend is approaching, a rational call holder may exercise early to own the shares in time to receive the dividend-especially when the remaining time value in the option is smaller than the dividend they'd collect. The covered call seller wakes up with shares called away sooner than expected.

This is not a reason to fear covered calls; it's a reason to operate them like a business and not a wish. Know the ex-dividend date for your holdings. Know whether your short call is in the money. Glance at the option's remaining time value when a dividend is near. Boring vigilance beats dramatic surprise.

What covered calls are great at (and what they are terrible at)

Covered calls shine in markets that are choppy, slightly bullish, or simply undecided-when prices wander, headlines rotate, and your patience is being taxed. The premium turns sideways time into productive time.

Covered calls are also excellent at *reducing the emotional volatility* of owning stocks. That's not a small benefit. A premium collected is a measurable outcome in a world full of maybes. It gives you a scoreboard you can actually control: you can choose strikes, expirations, position sizes, and you can choose whether to sell the call at all. You can't choose what the market does.

Where covered calls struggle is in explosive upside. When a stock gaps up and keeps going, a covered call can feel like you brought a bucket to a fire hydrant. You made money, but you didn't capture the full move. That's the cap showing up as opportunity cost.

They also struggle in severe drawdowns. The premium helps, but it's a thin cushion against a large fall. If the stock is down 30%, your 2% premium doesn't change the story; it just slightly improves the ending.

The practical takeaway is blunt: covered calls are not a substitute for stock selection and risk management. They are a way to monetize time and volatility on top of a stock position you already believe deserves capital.

A real-world way to think about ``success"

Many people judge covered calls incorrectly because they grade them like stock picking. They look back and ask, "Did I beat buy-and-hold?" Sometimes you will. Sometimes you won't-especially in roaring bull markets where caps are expensive.

A better grading system is operational:

- Did I sell calls only on positions I was willing to sell at the strike?
- Did the premium meaningfully reduce my cost basis over repeated cycles?
- Did the income improve my ability to stick with the plan during drawdowns?

- Did I avoid turning management into a stress hobby?

Covered calls are a production tool. You judge a production tool by reliability and process quality, not by whether it guessed the next headline.

Actionable takeaway: write one covered call like a professional, not a gambler

Before your next trade, do this on paper (or in a notes app) and refuse to skip steps:

1. Choose a stock (or ETF) you would be comfortable owning for the next year even if it drops materially.
2. Pick a strike where you would be genuinely happy to sell your shares. Compute your effective sale price: *strike + premium*.
3. Compute your covered-call break-even: *stock price paid – premium*. Write it down. That number is your buffer boundary.
4. Decide what you want more: more income (closer strike) or more upside room (farther strike). Make that choice consciously, not by impulse.
5. Place the trade and immediately set a calendar reminder a few days before expiration to decide whether you want to close, roll, or let it finish.

If you do nothing else, do that. One well-chosen covered call is enough to teach you the feel of the machine: cash arriving first, the stock doing what it does, and you learning to treat uncertainty not as an enemy—but as something the market will pay you for.

5.2 Cash-Secured Puts: Paid to Buy

A limit order is the most polite way to get ignored in the stock market.

You name your price, the market glances at it, and then—like a cat walking past an expensive toy—it does whatever it was going to do anyway. Days go by. The stock hovers a dollar above your bid as if it can smell your impatience. Meanwhile your cash sits there, behaving like a parked car: safe, inert, and quietly depreciating in real terms if inflation is doing its thing.

A cash-secured put is what happens when you stop asking politely and start charging admission.

You sell a put option-a contract that gives someone else the right to sell you shares at a specific price (the *strike*) by a specific date (the *expiration*). Because you're selling that right, you get paid up front (the *premium*). And because you're doing it *cash-secured,* you set aside enough cash to buy the shares if you're assigned. No borrowing. No improvising. If the trade turns into stock ownership, you can pay the bill the same day.

That's the core reframe: instead of placing a bid and hoping the market gifts you a fill, you sell an insurance policy on the stock and collect a premium for taking that risk.

The honest bargain: you're getting paid to accept an obligation

When you sell a put, you're making a promise. If the stock is below the strike at expiration (or if you're assigned earlier), you may have to buy 100 shares at the strike price per contract. That "100 shares" detail isn't trivia-it's the unit size of the whole machine. One put contract on a $50 stock is a commitment of about $5,000 of buying power (minus the premium you collected). One put contract on a $200 stock is about $20,000. The strategy is friendly; the math is not optional.

The premium is your compensation for carrying that obligation. It is also the market's opinion about how likely the obligation is to become inconvenient. Higher premium usually means higher uncertainty-earnings, headline risk, sector turbulence, or just a stock that likes to move like it had extra espresso.

This is why cash-secured puts feel so natural to disciplined investors: you only do them on stocks you'd be willing to own anyway. If you wouldn't want the shares at the strike, the premium is a bribe to do something you'll regret. Bribes don't age well.

A story you'll recognize: wanting the stock, hating the timing

Imagine you've been watching a company for months. You like the business, but the price feels rich. You don't need it in your portfolio today; you just don't want to miss it forever. So you set a mental "would buy" price-say $95 when it's trading at $100-and you wait.

The waiting is the problem. Idle cash is emotionally loud. It whispers that you're "doing nothing," that you're falling behind, that everyone else is making moves. A cash-secured put turns the wait into an action with a payoff.

Instead of placing a $95 limit order that pays you nothing, you sell the $95 put expiring in, say, 30 to 45 days and collect premium. Two things can happen, and both are legible outcomes:

- The stock stays above $95: the put expires worthless and you keep the premium. Your cash earned income while you waited.
- The stock falls below $95: you may be assigned and buy the shares at $95. You get the stock at your desired price (and your effective cost is lower because of the premium).

The trade isn't trying to win a debate about where the stock "should" go. It's monetizing your patience.

The payoff profile: a paid limit order with teeth

If you like clean mental models, this one is worth keeping:

A cash-secured put is a limit buy order that pays you a premium-but comes with the same downside as owning the stock from the strike price.

That last clause matters. If you sell a $95 put and the stock collapses to $60, you don't get to hide behind the premium. You're still obligated (if assigned) to buy at $95, and you'll be sitting on an immediate unrealized loss. The premium softens the first impact; it does not change the nature of gravity.

So what does it change?

It improves your entry price—if assignment happens. If you sell the $95 put for $2, your effective purchase price is $93 (strike minus premium). If you were happy buying at $95, being in at $93 is a better deal.

It pays you for "almost." If the stock never dips to your limit price, you still earned something. A traditional limit order pays you nothing for being "nearly right." The put premium is payment for standing ready.

It forces position sizing discipline—if you do it correctly. Because the obligation is explicit, you can't pretend a trade is "small" when it isn't. One contract is 100 shares. That's a feature, not a bug, as long as you respect it.

The cash part isn't a technicality-it's the safety system

"Cash-secured" is a promise you make to yourself before you make it to the market.

It means you reserve enough cash to buy the shares at the strike (typically strike × 100 per contract, offset slightly by the premium received). That cash is the collateral for your obligation. If the market drops and you're assigned, you can complete the purchase without scrambling, liquidating other positions, or using margin under stress—three behaviors that tend to turn a manageable drawdown into a portfolio-wide mess.

If you've ever seen someone blow up on "conservative" option income, it's rarely because the strategy was inherently reckless. It's because the person ran the machine faster than their balance sheet could feed it. They sold multiple puts across multiple tickers, told themselves "assignment won't happen on all of them," and then watched a broad market drop make that sentence feel naïve.

Cash-secured puts are boring in the best way: they make you pre-pay for your own responsibility.

Assignment: not a monster, just a delivery schedule

Assignment is the moment the market hands you the keys.

If you are assigned on a short put, you buy 100 shares per contract at the strike price. It can happen at expiration, and it can also happen before expiration because equity options are typically American-style (the holder can exercise any trading day). Early assignment on puts is generally less common than on calls around dividends, but it's not impossible—especially if a put is deep in the money and there's little time value left.

The practical implication is operational, not philosophical: if you sell a put, you should be prepared to own the shares at any point during the contract's life.

There's also the "paper cut" detail that surprises new sellers: at expiration, many brokers follow automatic exercise/assignment conventions when an option finishes even slightly in the money. You don't want to find out on Saturday morning that you bought 100 shares because the stock closed $0.03 below your strike on Friday. That's not a disaster if you sized it correctly. It *is* a disaster if you treated the position like a casual bet.

Choosing the underlying: you're underwriting a business, not just selling premium

When you sell a put, you're acting like an insurer. Insurers don't price policies by vibes; they ask what can go wrong, how badly, and how correlated it is with everything else in the book.

A good candidate for cash-secured puts has three qualities:

You actually want it. Not "it's popular," not "it has juicy premium," but "I'd be fine waking up with 100 shares at this strike." The option market can tempt you into writing puts on stocks you'd never buy outright. That's like selling homeowners insurance on a house you've never seen because the premium looked nice.

It trades options like a real marketplace. Liquid option chains matter. Tight bid/ask spreads mean you keep more of the premium rather than donating it to friction. If you can't get in and out without losing a noticeable chunk to spreads, you're running a factory where the conveyor belts are eating your product.

You can stomach the headlines. Stocks don't drop because the universe is mean; they drop because the future gets repriced. If you sell puts on a company whose earnings can be a coin flip, you're selling insurance on a hurricane zone and acting surprised when the weather app turns red.

This doesn't mean you must only trade the safest mega-caps. It means you must match the stock's personality to your own tolerance and your portfolio's capacity.

Strike selection: pick the price where you'd smile, not grimace

The strike is your "yes." It's the price at which you're agreeing to buy.

People often pick strikes the wrong way: they scan for a premium number that feels worth it, then back into whether they'd accept the shares. That's backward. First decide the price where you'd like to own the stock; then see what the market will pay you to make that commitment.

A useful way to sanity-check a strike is to calculate the *effective entry*:

$$\text{Effective Entry Price} = \text{Strike} - \text{Premium Received}$$

If the strike is \$95 and you collect \$2, your effective entry is \$93 (ignoring commissions and fees). That's the number to compare against your valuation, your chart level, or your "I can sleep at night" threshold.

Then ask a question that cuts through bravado: *If I already owned the shares, would I be comfortable holding them if they immediately dropped another 10%?* If that thought makes your stomach tighten, the strike is probably too close or the stock is too spicy for this role.

Expiration selection: don't sell time you can't manage

Selling puts is not hard; managing a portfolio of sold puts can become hard if every position expires on a different day and every week has a different "must-watch" event.

A calmer approach is to choose expirations that fit your attention budget. You want enough time for the premium to be meaningful, but not so much time that you forget why you sold it. Many traders like a consistent cadence-positions that mature on a predictable schedule-because it turns decision-making into routine.

Also consider event risk. If earnings are next week and you sell a put that expires right after earnings, you're not selling "time decay"; you're selling a lottery ticket to someone else and keeping the small part of the ticket. Sometimes that's fine if you priced it intentionally. Often it's an unforced error.

What cash-secured puts do to your psychology (and why that's valuable)

The market is a machine that manufactures regret. Buy now and it dips; wait and it runs. Cash-secured puts are one of the few tools that consistently reduce that regret because they let you win in multiple ways.

If the stock rises, you didn't "miss out" entirely-you earned premium. If the stock falls to your strike, you didn't "catch a falling knife" at market price-you bought at a pre-agreed level with a premium offset. If the stock chops sideways, you got paid for boredom-a rare perk in modern markets.

This doesn't remove risk. It removes the feeling that you must be clairvoyant to be successful. You're replacing prediction with process.

The two classic mistakes: pretending collateral is optional and chasing premium

Most costly errors in cash-secured puts come from two temptations:

Treating "cash-secured" as a suggestion. If you don't actually reserve the cash-if you sell multiple puts using margin or by assuming you can "handle it" later-you've changed the strategy into something else entirely. You've turned a paid bid into leveraged exposure. It might work for a while, until it fails in a way that forces you to sell at the worst time.

Letting premium pick your stocks. The market will happily pay you more to insure riskier names. That's not generosity; it's pricing. If you chase the fattest premiums, you're selecting for the companies most likely to give you assignment at exactly the moment you least want it.

There's nothing wrong with occasionally selling puts on higher-volatility names *if* your position size is appropriate and you truly want the shares. The mistake is doing it because the premium felt exciting.

Excitement is not an edge. It's usually a fee.

A practical "factory" workflow: how to run one put sale end-to-end

A cash-secured put works best when it has a written intent and a pre-accepted outcome. Here's a clean workflow that keeps it grounded:

1. Pick the stock and write a one-sentence reason you'd like to own it (business quality, valuation, long-term role in portfolio).
2. Choose the maximum number of shares you'd be happy owning. Convert that into contracts (shares / 100).
3. Set aside the collateral *before* selling the put: strike × 100 × contracts, minus premium as a small offset.
4. Choose a strike that represents a price you'd genuinely accept. Compute effective entry (strike – premium).
5. Decide in advance what you'll do if assigned: keep the shares as a long-term holding, or treat it as inventory for income generation.

That last step is more powerful than it sounds. If you already know how you'll behave after assignment, assignment stops being a crisis and becomes a schedule.

Actionable takeaway: place one "paid bid" with guardrails

Do this the next time you're tempted to place a limit order:

1. Identify a stock you would be happy to own at a lower price.
2. Pick a strike at that price and an expiration far enough out that the premium feels meaningful *without* crossing a major binary event you haven't priced (like an imminent earnings report).
3. Sell *one* cash-secured put contract and reserve the full collateral in cash.

4. Immediately write down two numbers on a sticky note (digital or physical): *effective entry* (strike − premium) and *total collateral committed* (strike × 100).

If the put expires worthless, you just got paid for patience. If you're assigned, you just bought a stock you wanted at a discount. Either way, you turned idle cash into a working asset-without pretending risk doesn't exist.

5.3 The Wheel Strategy

The Wheel is what happens when you stop treating options trades like one-off stunts and start running them like a shift schedule.

Most investors already live in a loop, whether they admit it or not: cash becomes shares, shares become cash, and the cycle repeats whenever fear or greed hits a satisfying volume. The Wheel takes that messy, emotional loop and replaces it with a deliberate one—two contracts, two obligations, one repeating workflow that tries to turn time into a paycheck.

It's also the strategy most likely to be oversold as "easy income." That's a shame, because the Wheel doesn't need hype. It needs respect. Run it with good underlying selection and sane sizing, and it can feel like owning a small toll road. Run it on the wrong stock or with too many contracts, and it can feel like you volunteered to catch falling knives for tips.

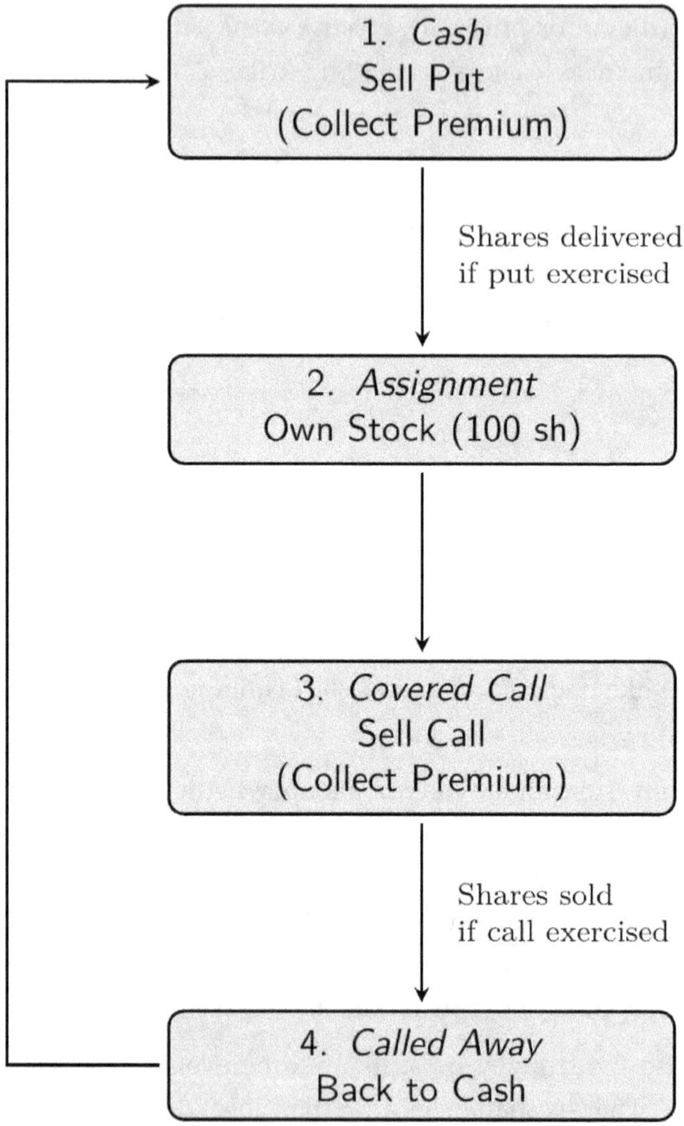

There are only two moving parts here—selling puts and selling calls—but the loop changes your posture. You're no longer staring at a position wondering when you should act. You already know what you do when you're in cash, and you already know what you do when you're holding shares. The Wheel gives each state a job.

The two obligations that make the machine run

The Wheel is built on two promises you make to the market:

- When you sell a put, you're promising to *buy* shares at the strike if assigned.
- When you sell a covered call, you're promising to *sell* shares at the strike if assigned.

That's it. There's no hidden third lever. The strategy is a loop of "paid bids" and "paid offers."

The elegance is that both promises are paid up front. The market hands you premium for being willing to transact at specific prices. The danger is that both promises are binding. If the stock moves hard against you, you don't get to pretend you were just "selling premium." You were underwriting a purchase price, and then underwriting a sale price.

Why the Wheel feels so good when it's working

A well-run Wheel scratches a deep itch: it makes you feel *useful* in the market.

When you're in cash, you're not waiting—you're getting paid to name your buy price. When you're in shares, you're not hoping—you're getting paid to name your sell price. The cycle gives you something to do that isn't the usual doom-scrolling of quotes.

It also creates a surprisingly healthy form of detachment. You stop needing every trade to be a heroic win because the goal isn't the perfect entry or the perfect exit. The goal is repeatable cash flow with acceptable outcomes. That mindset is rare in a world where investing advice is often a contest of who can sound most certain.

A concrete example: one lap around the Wheel

Suppose a stock is trading at $50 and you'd be happy owning it around $48. You sell a cash-secured put:

- Sell 1 put contract, strike $48, expiration 30–45 days out
- Collect $1.00 premium ($100 total)
- Set aside $4,800 collateral (strike × 100)

Now the fork in the road:

If the stock stays above $48, the put expires worthless. You keep $100 and you're still in cash. You can sell another put and keep the loop moving.

If the stock falls below $48 and you're assigned, you buy 100 shares at $48. Because you collected $1.00 premium, your effective cost basis is $47.

Once you own shares, the Wheel shifts from "paid bid" to "paid offer." You sell a covered call:

- Sell 1 call contract against your 100 shares, strike $50, expiration 30–45 days out

- Collect, say, $1.10 premium ($110 total)

Now you have a new fork:

If the stock stays below $50, the call expires worthless. You keep $110 and you still own the shares. You can sell another call next cycle.

If the stock rises above $50 and you're assigned, your shares are sold at $50. Your share profit is roughly $3 per share (sold at $50, effective basis $47), plus you collected premiums along the way. You're back in cash, which is exactly where the Wheel likes to be after a profitable sale.

Notice what you *didn't* have to do: guess the future. You simply agreed on prices you could live with and got paid to make those agreements.

The Wheel's psychological trick: assignment is a phase shift, not a failure

Most option sellers fear assignment the way people fear surprise bills. The Wheel treats assignment like a shipment arriving at a warehouse: it changes what you do next, but it isn't an emergency.

That shift matters because fear creates expensive decisions. A trader who views assignment as "getting stuck" tends to overpay to avoid it—buying back options at bad prices, rolling out of panic, or picking strikes that are so far away that the premium turns into spare change. A trader who views assignment as a planned phase tends to make calmer choices: if shares arrive, start selling calls; if shares leave, start selling puts.

This is not a motivational poster. It's operational reality. The Wheel is easiest when you stop moralizing outcomes and start running the checklist.

The hidden risk: the Wheel is still long stock risk, sometimes at the worst moment

Here's the sentence that keeps the strategy honest:

If the stock falls hard, the Wheel can concentrate you into share ownership right when you least want it.

That's not a bug; it's the nature of selling puts. When markets are calm, selling puts can feel like being paid for patience. When markets are violent, selling puts can feel like being paid to stand in front of a fast-moving door.

If you sell a put and the stock drops through your strike, you may be assigned and become a shareholder during a drawdown. Then you sell calls, which cap your upside during the rebound. That combination—downside exposure during the fall, capped upside during the rise—is why the Wheel can lag a roaring bull market after a sharp correction.

The premium you collect is the compensation for accepting that path. If you collect small premiums on a stock with big downside potential, you're selling hurricane insurance for a fair-weather price.

The real edge: turning volatility into a schedule

The Wheel's best use case is not "beat the market." It's "turn market movement into a manageable routine."

Volatility is usually framed as something that happens *to* you. The Wheel treats volatility as something you can *invoice*. When implied volatility is higher, premiums are richer. When implied volatility is lower, premiums are thinner and the strategy naturally cools down.

That doesn't mean you blindly sell puts when volatility spikes. It means you recognize why the strategy exists: the market pays you to take on uncertainty. The Wheel is a way to package that uncertainty into repeatable actions.

Choosing the right underlying: the Wheel is a long-term relationship

If you pick the wrong stock, the Wheel will make sure you spend a lot of time with it.

That's not a joke. When a stock trends down for months, you can end up holding it and selling calls below your original entry—collecting premiums while watching your capital remain tied up. Sometimes that's acceptable; often it's miserable.

So the best Wheel candidates are not necessarily the ones with the highest premium. They're the ones you would be willing to own through a rough season without constantly renegotiating your own conviction. Think durable businesses and broad ETFs with deep options liquidity—underlyings that won't turn into a permanent repair job the moment the market mood changes.

Liquidity deserves special respect. Wide bid/ask spreads are like a tax you pay every time you touch the machine. The

Wheel involves repeated trades. A small "spread tax" compounds into real money over time.

Position sizing: one contract is a real amount of risk

The Wheel's smallest unit is 100 shares. That's convenient for options settlement and annoying for portfolio construction.

A single contract on a $150 stock is $15,000 of notional exposure if assigned. That's not a "toe in the water" for many accounts. The danger is that people scale the Wheel by opening too many positions because each premium looks small in isolation—$120 here, $180 there—until assignment turns those small credits into large concentrated holdings.

A practical guardrail: decide how much of your portfolio you're willing to allocate to any single underlying *before* you sell the put. Then translate that into the maximum number of contracts. If your limit is 5% per position and your portfolio is $200,000, your maximum exposure is $10,000. That means one contract on a $90 strike put ($9,000) might be fine; one contract on a $130 strike put ($13,000) violates your own rules before the first premium hits your account.

The Wheel rewards the investor who respects math more than mood.

When to keep spinning, when to step off

A common misconception is that the Wheel must be continuous—that you must sell the next option as soon as the last one expires. That's how you turn a strategy into a compulsion.

Sometimes the right move is to pause:

- If a stock's fundamentals change, you don't owe it loyalty just because you sold a put.
- If implied volatility collapses and premiums become thin, the compensation may no longer justify the obligation.
- If your portfolio is already heavy in one sector, selling more puts there can quietly load correlation risk.

Stepping off the Wheel isn't failure. It's risk management. A factory that never shuts down for maintenance eventually shuts down for reasons it didn't choose.

The quiet superpower: the Wheel can train better investor behavior

There's a behavioral reason the Wheel remains popular: it nudges you toward buying and selling at pre-committed prices, with cash flow as a reward for discipline.

Selling puts encourages you to buy on weakness (at a discount you pre-selected) instead of chasing strength. Selling covered calls encourages you to take profits at a pre-selected level instead of waiting for "just a little more." Those are both behaviors that sound easy in theory and are strangely hard in the moment.

The Wheel doesn't remove emotion, but it can sandbox it. You still feel the market's drama—you're human—but you also have a script, and scripts are calming when the crowd is loud.

Actionable takeaway: build a ``Wheel-ready" checklist and run one small cycle

Before you try to spin the Wheel across ten tickers, run one controlled lap with guardrails:

1. Pick one underlying with highly liquid options and a business (or index) you'd be comfortable owning for months if needed.
2. Set a hard maximum allocation for that one underlying (as a percentage of portfolio). Convert it into the maximum number of contracts.
3. Choose a put strike where you would be happy buying 100 shares, and reserve the full cash collateral before placing the trade.
4. If assigned, immediately switch to selling a covered call at a strike where you would be happy selling the shares. If not assigned, decide whether you still want to keep selling puts *based on today's price and today's premium*, not because you're on autopilot.

If you can do that calmly—without bargaining with yourself on strike selection, without "accidentally" oversizing, and without treating assignment like a personal insult—you've built something rare: a repeatable cash-flow machine that doesn't require you to be right about tomorrow.

5.4 Credit Spreads for Defined Risk

Some stocks feel like they were priced for a different species.

You open the quote screen and there it is: $600, $900, $1,200 a share—numbers that make the standard "100 shares per options contract" rule look less like a convention and more like a velvet rope. Even if you *could* buy 100 shares, you might not want to. Concentrating that much capital into one name can turn a normal pullback into a personal crisis.

Credit spreads exist for that exact moment. They let you sell option premium on big-ticket underlyings—or simply keep risk small on any underlying—without pretending you have to own the whole building to collect rent from one unit.

A credit spread is a two-option position: you sell an option to collect premium, and you buy another option farther out-of-the-way as protection. The result is income up front with a loss that is capped from the start. In other words: you get paid to take risk, but you also buy a fire extinguisher before you light the stove.

Why ``defined risk" is more than a comforting phrase

Selling a cash-secured put is honest risk: you're paid, and if the stock collapses you can end up owning shares far above market price. That's not a flaw; it's the cost of being willing to buy. The catch is that the worst-case dollar drawdown can be large because stock downside is large.

A defined-risk spread changes the contract you're writing. You still sell insurance to someone else, but you also buy insurance for yourself. You are deliberately giving up some premium to eliminate tail risk—the kind of "it can't happen" move that has a way of happening at the worst possible time.

That tradeoff is the heart of vertical spreads:

- Less income than selling a naked put.
- Less risk than being forced into stock ownership through a deep drawdown.

If you've ever slept badly with an oversized position, "less risk" is not a footnote. It's the point.

The basic structure: a bull put spread (credit put spread)

The most common income-oriented credit spread is the *bull put spread,* also called a *credit put spread*. It's built from two puts with the same expiration date:

- *Sell* a put at a higher strike (closer to the current stock price).
- *Buy* a put at a lower strike (farther from the current stock price).

Because the put you sell is typically more expensive than the put you buy, you receive a *net credit* up front. That net credit is the most you can make.

The put you buy is your seatbelt. If the stock plunges, that long put rises in value and limits how much you can lose. You are not hoping the market is kind; you are buying a contract that makes cruelty affordable.

A concrete example: income with guardrails

Say a stock is trading at $200. You'd like to earn income by betting it stays above $190 over the next month, but you don't want the obligation of buying 100 shares at $190 (a $19,000 commitment) if the market turns ugly.

So you do this instead:

- Sell the $190 put.
- Buy the $185 put.
- Same expiration for both.

Suppose you collect $2.00 for the short $190 put and pay $1.20 for the long $185 put. Your net credit is $0.80 per share, or $80 per spread.

Now the defining characteristics become wonderfully clear:

Max profit = net credit received = $80.

If the stock stays above $190 at expiration, both puts expire worthless. You keep the $80. No shares change hands. No drama.

Max loss = spread width – credit.

The spread width is $5 (190–185). Width × 100 = $500. Subtract the $80 credit you already received: max loss = $420 per spread.

That's the "defined risk" promise in dollars. Before you place the trade, you know the worst-case loss on that one spread is $420, not $19,000 of forced share ownership. The market can still humble you, but it can't bankrupt you through this single position unless you deliberately oversize it.

Breakeven = short strike – credit.

Breakeven here is $190 – $0.80 = $189.20. If the stock finishes above $189.20 at expiration, you're profitable; below it, you're losing (up to the capped max loss).

That's a lot of clarity for a two-leg trade.

What you're really trading: probability vs. payout

Credit spreads are often described as "high probability, low reward." That's not always accurate, but it points at a real structural feature: you're collecting a relatively small credit in exchange for taking a relatively larger defined risk.

If you make $80 while risking $420, your return on risk is about 19% for that cycle (before commissions/fees). That can be attractive—but notice the shape: you may win many times in a row and then give back a chunk of those gains in one bad month if you're careless about underlying selection or if you size too aggressively.

This is where credit spreads separate mature traders from adrenaline traders. The goal is not to win 95 times and then pretend the 96th loss was an unfair glitch. The goal is to build a playbook where the occasional loss is expected, affordable, and sized so it doesn't knock the whole operation off its feet.

Why spreads help with expensive stocks

A cash-secured put on a $500 stock ties up around $50,000 of collateral per contract. That's not "portfolio management" for most people; that's "one decision now runs my whole year."

A credit spread lets you express a similar idea—"I think this stock stays above a certain level"—with a fraction of the capital at risk. Your broker will typically require margin roughly aligned with the maximum loss of the spread, not the full notional value of 100 shares at the strike. That means you can participate in liquid option markets without either (a) being rich enough to ignore concentration or (b) doing mental gymnastics to justify it.

The spread is also a psychological relief. When the market drops, you don't have to wonder whether you're about to be forced into a huge stock purchase. You already know the most you can lose. That knowledge changes the quality of your decision-making under pressure.

The trade's personality: you are selling a range, not a direction

A covered call feels like "I'm willing to sell above here." A cash-secured put feels like "I'm willing to buy below here." A credit spread is subtler: it's closer to "I think the stock stays on this side of a boundary long enough for time to decay."

For a bull put spread, the boundary is the short strike. You don't need the stock to soar. You need it to *not break down*. This makes credit spreads naturally suited to underlyings that are choppy but not collapsing, or to situations where you like the business but don't want to deploy full share capital.

It also means spreads can tempt you into complacency. The P/L often looks fine...until it isn't. Which brings us to the part that deserves real respect.

The operational reality: a spread is defined-risk, not no-risk

The long put defines your maximum loss at expiration. That's a crucial difference from an uncovered short put. But several real-world mechanics still matter:

Early assignment can still happen. The short put you sold is a standalone contract with its own exercise rights for the holder. Assignment before expiration is possible. If it happens, you can suddenly find yourself owning shares (long 100) while still holding the long protective put. That position is not "broken," but it requires you to act: you may need to sell the shares, exercise/sell the long put, or close the whole structure. Defined risk doesn't mean "ignore your account for 45 days."

Pin risk near expiration is real. If the stock hovers around your short strike into expiration, small price moves can decide whether you're assigned. Many brokers follow automatic exercise conventions for options that finish even slightly in the money. The practical lesson is simple: if a spread is near the money late on expiration day, pay attention. The market loves creating tiny, annoying surprises.

Liquidity and bid/ask spreads matter twice. You have two legs. If the options are illiquid, you pay the spread cost on the way in and the way out, and it can chew up a large portion of your credit. Spreads are not forgiving of sloppy fills.

None of this is meant to scare you off. It's meant to keep the strategy in the adult category: controllable, but not automatic.

Credit spreads as ``renting with a deductible"

Here's a mental model that sticks:

Selling a naked put is like insuring someone else's house against fire-if the whole place burns down, you may be on the hook for a huge rebuild.

A credit spread is like selling that insurance *but also buying reinsurance* for yourself. You still collect a premium, but you've built in a deductible and a limit. Your long option is the contract that says: "Past this point, I'm no longer the one absorbing the catastrophe."

That's why spreads are so popular among traders who care about survival. If you can't survive the bad months, the good months don't matter.

Making spreads practical: the three numbers you must calculate every time

Before placing any credit spread, write down these three numbers. If you can't write them down quickly, you don't understand the trade yet—and that's not a moral failing, it's just a sign to slow down.

- *Max profit* = net credit ×100
- *Max loss* = (strike width ×100) − max profit
- *Breakeven* = short strike − net credit

That short list prevents a long list of mistakes.

To make it easy to reference, here's a compact template you can keep nearby.

Item	*Formula (Bull Put Credit Spread)*
Net credit (per share)	Short put premium – long put premium
Max profit (per spread)	Net credit ×100
Spread width (per share)	Short strike – long strike
Max loss (per spread)	(Spread width ×100) – Max profit
Breakeven (per share)	Short strike – net credit

The table is not busywork. It's a way to keep your brain from turning "defined risk" into "I don't need to think."

Why spreads demand humility about sizing

Spreads look small. $80 credit. $420 max loss. Those numbers feel contained, which can encourage you to stack too many of them. Ten spreads suddenly means $4,200 of max loss on one ticker. Twenty spreads means $8,400. If a broad market selloff hits and multiple positions move against you at once, "defined risk" can still become a defined headache.

The real sizing question is not, "What's the premium?" It's, "If I took the max loss on this position, would my portfolio still be running smoothly tomorrow?" If the honest answer is no, the position is too large. Income strategies are supposed to make your finances calmer, not turn your evenings into improvisational math.

A word on expectations: spreads are not a substitute for owning great assets

Credit spreads are a tool for harvesting time decay and volatility risk premia. They are not a magic replacement for long-term compounding from great businesses. Think of them like a side business: done well, they create cash flow and improve capital efficiency; done poorly, they distract you from the core job of owning assets that appreciate.

That's why many investors use spreads selectively:

- to generate income on underlyings they don't want to own in 100-share blocks,
- to keep risk small while learning,
- to participate in premium selling when cash-secured puts would tie up too much capital.

The strategy is a toolbelt, not a religion.

Actionable takeaway: place one ``training-wheels" credit spread with pre-written exits

Here's a practical way to start without turning it into a science project:

1. Choose a highly liquid underlying with tight option spreads.
2. Pick a bull put spread with a small width (e.g., $5 wide) so the max loss per spread is easy to stomach.
3. Calculate and write down: net credit, max profit, max loss, and breakeven.

4. Decide your management rule *before* entering. Two common ones: take profits early when most of the credit is earned, or cut risk if the short strike is clearly threatened.
5. Enter only one spread for your first cycle, and treat the goal as learning execution and behavior, not squeezing every dollar.

If you can execute one small spread cleanly—good fills, known risk, calm management—you've added a powerful tool to your income factory: premium collection with guardrails, even when the underlying is too expensive or too concentrated to own outright.

5.5 Selecting Strikes and Expirations

Choosing a strike price and an expiration date looks like a tiny decision—two dropdown menus, a couple of clicks. In practice it's the moment you decide what kind of income operator you are: the patient one who collects steady rent, or the frantic one who keeps rewriting the lease every time the market clears its throat.

Options aren't vague. They're contracts with hard edges. The strike is the price you're agreeing to transact at if the market pushes you there. The expiration is the calendar deadline on your commitment. Everything else—premium, probability, stress level, how often you'll need to babysit trades—cascades from those two settings.

A good framework starts with a brutally simple question: *What job is this trade doing in my portfolio?* If you don't know the job, you'll be tempted to optimize for whatever number looks best on the screen today, which is like hiring a contractor because they smiled at you.

The strike is your price. The expiration is your clock.

Strikes and expirations are dials, not decorations.

Strike: where the deal happens. When you sell a cash-secured put, the strike is your agreed purchase price. When you sell a covered call, the strike is your agreed sale price. When you sell a credit spread, the short strike is your boundary line (the level you're betting won't be breached), and the long strike is your seatbelt.

So the strike is not merely "how much premium can I get?" It is a statement of willingness:

- *Covered call:* "I'm willing to sell my shares at $\$K$."
- *Cash-secured put:* "I'm willing to buy shares at $\$K$."
- *Credit spread:* "I'm willing to risk $\$X$ that the stock stays on this side of $\$K$."

Expiration: how long you're on the hook. Time is a sneaky kind of risk. The longer you give a contract to live, the more opportunities the world has to surprise you: earnings, downgrades, lawsuits, sector rotations, recessions, "oops" product announcements, geopolitics, and the market's favorite hobby—overreacting.

Expiration is also how you sell *time value*. Options contain value that exists purely because time remains before the contract ends. As time passes, that value tends to decay, which is why selling options can generate income even if the stock doesn't move much. You're not just betting on direction; you're selling a melting ice cube to someone who hopes it doesn't melt too fast.

The two risk gauges worth knowing: Delta and Theta (without the calculus)

Options traders love Greek letters the way chefs love fancy knives: useful, but often waved around to impress people at parties. You don't need a PhD to use the two Greeks that do most of the practical work in income strategies.

Delta: the "moneyness" and rough odds gauge. Delta is commonly treated as a rough proxy for how likely an option is to finish in the money by expiration. For income sellers, it's a quick way to describe how aggressive your strike choice is.

Think of it like this:

- A lower absolute delta (like 0.15) usually means a farther-out strike: lower premium, lower chance of assignment.
- A higher absolute delta (like 0.40) usually means a closer-to-the-money strike: higher premium, higher chance of assignment.

Delta isn't a promise. Markets don't sign guarantees. But delta is a useful *dial* because it keeps you from choosing strikes based purely on vibes.

Theta: the "time decay" meter. Theta estimates how much option value decays per day from the passage of time, all else equal. When you sell options, you generally want time decay to work in your favor. Theta is the engine behind the "paid for patience" feeling: as days pass, the option you sold tends to lose value, and you can often buy it back cheaper (or let it expire).

Here's the key intuition: *time decay is not linear*. The ice cube melts faster near the end.

Why many income traders like 30--45 days to expiration

There's a sweet spot where you're selling a meaningful amount of time value without marrying the contract for too long. Many income-focused traders find it in the neighborhood of 30–45 days to expiration (often written as 30–45 DTE).

At very short expirations, premium can evaporate quickly, but the position can behave like a hummingbird—small, twitchy, and easily spooked by minor price moves. At very long expirations, you may collect more premium, but you also lock in your obligation for a long time, and your capital (or shares) can feel tied to a slow-moving decision.

The most important reason 30–45 DTE keeps showing up is the shape of time decay. As expiration gets closer, the option's remaining time value often erodes faster—creating an "acceleration zone" where the income engine tends to rev higher.

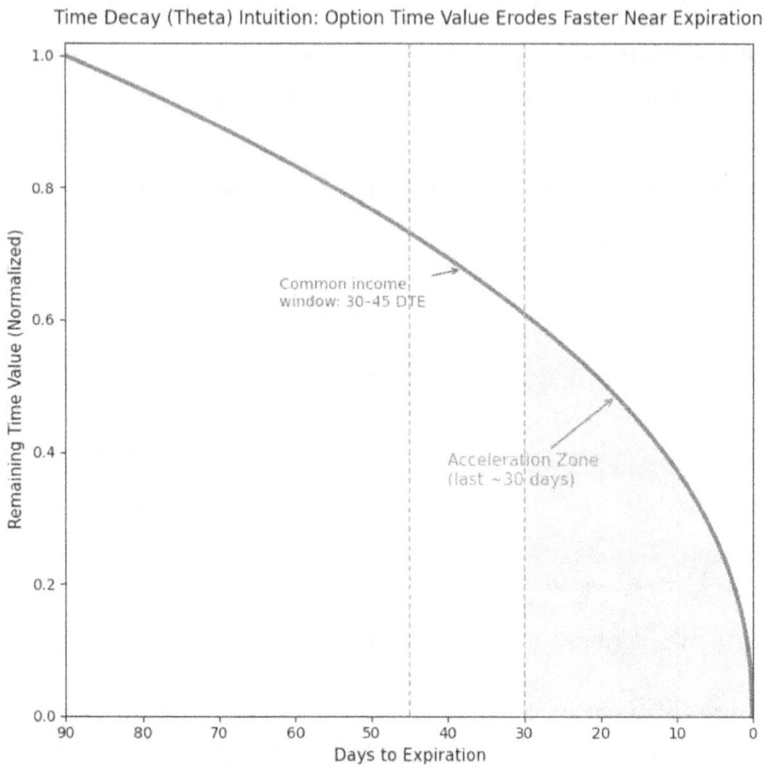

That curve is why 30–45 DTE often feels like a practical "production line" setting: you're stepping into the part of the calendar where decay tends to accelerate, while still leaving enough time to manage the position without everything turning into a last-minute scramble.

Strike selection as a trade-off between rent and regret

Premium is seductive because it's immediate and measurable. Opportunity cost is sneaky because it's hypothetical until the day it punches you in the face.

Covered calls: the closer strike pays more, but sells your upside cheaper. If you sell a covered call close to the current price, you usually collect more premium. You also increase the odds your shares get called away—potentially right before a big rally. That's not "bad" if you chose a strike you'd happily sell at. It's bad if you chose it because the premium looked tasty and you weren't emotionally prepared to lose the shares.

A useful discipline: always compute the *effective sale price*:

$$\text{Effective Sale Price} = \text{Call Strike} + \text{Premium Received.}$$

Then ask: *Would I place a limit order to sell my shares at this effective sale price within this timeframe?* If the honest answer is no, the strike is too low for your goals.

Cash-secured puts: the closer strike pays more, but buys the stock sooner (and higher). Selling puts closer to the current stock price often boosts premium, but it also means you're agreeing to buy at a higher price and with a higher chance of assignment. That can be perfectly rational if you want the stock soon. It can also be a way of accidentally catching the falling knife earlier than you intended.

A matching discipline: compute the *effective entry price*:

$$\text{Effective Entry Price} = \text{Put Strike} - \text{Premium Received.}$$

If that number doesn't look like a price you'd happily pay, step back. The premium is not a coupon; it's compensation for taking a real obligation.

Credit spreads: the short strike is the cliff; the long strike is the parachute. With spreads, people obsess over the credit and forget the geometry. A tighter spread (small width) limits max loss but also limits the credit you can collect. A wider spread increases risk and often increases credit, but not always proportionally. Your job is to decide what you want to be true: "small, repeatable" or "bigger, less forgiving."

A practical ``Goldilocks" recipe: 30--45 DTE and about 0.30 delta

If you want a starting point that many income operators find workable, this is a common one:

- Choose an expiration around *30–45 days*.
- Choose a strike around *0.30 delta* (approximately) for the short option.

Why does this combination show up so often?

Because it balances three competing forces that otherwise fight:

- Premium that's meaningful enough to matter,
- strikes far enough away to give you breathing room,
- and an expiration window where time decay is starting to get serious.

It's not a law of nature. It's a sensible default. If you always start with a default, you'll make intentional adjustments rather than random ones.

Static return vs. if-called return: two ways to stop lying to yourself

Income trading has a classic self-deception: counting premium as "return" without acknowledging what you gave up or what you committed.

Two quick calculations keep you honest, especially on covered calls:

Static return (premium yield). How much premium did you collect relative to the stock value or collateral, over the time period?

For a covered call:

$$\text{Static Return} \approx \frac{\text{Premium}}{\text{Stock Price}}.$$

For a cash-secured put:

$$\text{Static Return} \approx \frac{\text{Premium}}{\text{Cash Collateral}}.$$

This is the "rent check" view. Useful—but incomplete.

If-called (or if-assigned) return. On a covered call, the "best-case" outcome is often being called away at the strike. That outcome includes premium plus any share gain up to the strike:

$$\text{Max Gain per Share} = (\text{Strike} - \text{Entry Price}) + \text{Premium}.$$

This is the "what did I agree to sell for?" view. It forces you to price your cap.

Neither number is the truth by itself. Together, they keep you from optimizing for a single shiny metric.

Event risk: the calendar is part of the price

Markets don't move randomly; they move around events. Earnings, dividends, economic releases, FDA decisions, court rulings, product launches—these are scheduled cliffs where implied volatility tends to inflate.

You can either treat that as an opportunity or a hazard, but you can't pretend it doesn't exist.

If you sell options through earnings, you are selling a jump. The premium may be higher for a reason: the market is pricing the possibility of a sudden move. If your strategy depends on gentle time decay, selling right into a known volatility event can feel like renting out your house to someone hosting a party. They paid extra. There's a reason.

If you sell calls near ex-dividend dates, assignment risk can change. Covered calls can be assigned early, and dividend timing can influence that risk when calls are in the money and there's little time value left. You don't need to obsess over it daily, but you do want the habit of checking the calendar when a dividend is near and your short call is deep in the money.

The ``attention budget'' rule: choose expirations you can manage well

A secret constraint in options income is not volatility. It's you.

If you sell weekly options because the premium looks efficient, but you can't reliably monitor positions, you'll eventually miss something important and pay for it. If you sell very long-dated options because you don't want to think about them, you may lock your portfolio into caps and obligations that no longer match reality.

A sustainable cadence is one you can execute consistently:

- consistent expirations (so you're not constantly re-learning your own portfolio),
- position sizes that don't demand heroics,
- and strikes chosen so that assignment is an acceptable outcome, not an existential threat.

Consistency is underrated alpha.

A compact checklist you can actually use

Here's a practical decision path that works across covered calls, cash-secured puts, and credit spreads:

1. *Define the acceptable assignment outcome first.* "If assigned, I'm okay buying 100 shares at $\$K$" or "If called away, I'm happy selling at $\$K$."
2. *Choose a default expiration window* (often 30–45 DTE) that fits your schedule and your desire for time decay.
3. *Use delta as a strike compass.* Start around 0.30 for a balanced stance; move lower for more conservatism, higher for more income and more assignment likelihood.
4. *Check the calendar for known events* (earnings, dividends) and decide whether the extra premium is worth the extra jump risk.

5. *Write down the two numbers that matter:* effective entry/sale price, and worst-case exposure (shares or max loss).

None of this requires forecasting. It requires honesty.

Actionable takeaway: set your ``house settings" and stop improvising

Pick one underlying you trade often (an ETF or a highly liquid stock) and establish your default settings for the next month:

- A default expiration window: choose either 30–45 DTE (common for income) or a different range that matches your attention budget.
- A default delta target for short options: start at 0.30.
- A rule for when you will *not* sell options (for example, immediately before an earnings report unless you're intentionally selling that risk).

Then place your next trade using those defaults *unless you can clearly explain why you're deviating*. That single discipline—default first, deviation by reason—turns strike and expiration selection from random clicking into a repeatable craft.

5.6 Managing Positions: Rolling and Defense

The market has a particular sense of humor: it waits until you feel "safe" and then reminds you why premiums exist.

If you sell options for income, you are volunteering to be the adult in the room when prices swing. That doesn't mean you'll always be right. It means you'll always have a plan. Without a plan, a short option position can turn from "cash flow" into "why am I awake at 2 a.m. staring at a chart." With a plan, it becomes what it should be: a managed obligation with known exits.

Position management is where option selling stops being a clever idea and becomes an actual craft. The entry is a promise; management is how you keep that promise from quietly wrecking your portfolio.

First principle: treat every adjustment as a brand-new trade

Rolling and defending can feel like you're "fixing" something. That framing is dangerous because it invites denial: you cling to the original thesis, the original strike, the original ego.

A healthier framing is brutally practical: when you roll, you are closing one position and opening another. The market doesn't care that you call it an adjustment. Your account statement won't either. So every time you consider a roll, ask a question that's slightly uncomfortable on purpose:

> If I had no position right now, would I initiate *this new one* at these strikes, at this expiration, for this credit?

If the answer is no, rolling is not "defense." It's procrastination with extra steps.

The only three reasons to manage a short option

Most management decisions fall into one of three motives. If you can't name the motive, you're probably reacting, not managing.

1) **Lock in a win and free up capital.** Income traders often discover a weird truth: the last 20% of potential profit can be the most dangerous 20%. If you sold an option for $1.00 and it's now worth $0.20, most of the decay has already happened. What's left is a small reward for keeping risk on the table.

Closing early can be boring—and boredom is often the smell of good risk management. You book the profit and redeploy capital into a fresh setup with more time value to harvest.

2) **Reduce risk when the strike is being threatened.** When the stock drifts toward your short strike, the position changes personality. What used to be a calm bet on "nothing dramatic happens" becomes a high-sensitivity position where every dollar move starts to matter.

That is the moment to decide whether you still want the same obligation at the same price. Sometimes the answer is yes: you accept assignment on a put, or you accept being called away on a covered call. Other times the answer is no: you'd rather move the obligation farther away in exchange for time or reduced credit.

3) **Exit because the thesis is broken.** This is the one people avoid because it hurts. Maybe the stock had a fundamental change, maybe the sector cracked, maybe volatility spiked for a reason you can't underwrite. A disciplined factory shuts down

a malfunctioning line instead of feeding it more raw material and hoping it behaves.

If your original reason for trading the underlying no longer holds, "rolling" is not brave. It's stubborn.

Know your exposures: stock risk versus option risk

A short option position can look like "just an option," but it often behaves like a stock position in disguise.

Covered calls are still long stock. If the stock drops hard, the premium is a small cushion, not a shield. Managing a covered call under pressure is usually about managing the stock exposure, not the option.

Cash-secured puts are stock exposure waiting to happen. When a put strike is threatened, you're not merely defending a premium trade-you're deciding whether you want to buy shares at that price, soon.

Credit spreads are defined-risk, but still sensitive. Spreads cap the worst-case loss at expiration, but near the short strike they can move fast and widen spreads can make exits expensive. Defined risk doesn't mean "no management required." It means your maximum pain has a ceiling.

Once you see the true exposure, the right action becomes clearer. People get into trouble when they manage the option while ignoring the exposure underneath it.

The 50% rule: a boring habit that saves real money

Many income traders follow a simple management rule: close a short option position when about 50% of the maximum profit is achieved.

If you sold an option for $1.00, buy it back when it's around $0.50 (or close the spread when you can keep roughly half the credit). The logic is not that 50% is sacred; the logic is that risk doesn't decline linearly with remaining profit.

When you've captured half the credit early in the cycle, you often have:

- less time left for the market to surprise you,
- less premium left to earn,
- and roughly the same headline risk as always.

Closing at 50% turns "income" into realized income. It also shortens your average time in trades and reduces the number of positions you carry into expiration week—the week where little mistakes become real surprises.

The best part: the rule is emotionally neutral. It doesn't require you to guess direction. It requires you to keep promises to yourself.

Rolling: what it is in plain English

Rolling is the act of:

- buying back the option you sold (closing the original obligation), and

- selling another option (opening a new obligation), usually with a later expiration, sometimes with a different strike.

You can roll "out" (same strike, later expiration), "up" or "down" (different strike), or both. For covered calls, you might roll up and out to raise the sale price and give the stock more time to behave. For puts, you might roll down and out to lower your potential entry price while buying time.

Here's the key operational truth: *rolling is not free*. A roll is a trade. It has a price (debit or credit), and it changes your exposure.

If you roll for a credit, you're often being paid to extend your obligation. If you roll for a debit, you're often paying to change the shape of your risk—usually to reduce assignment likelihood or to move the strike away from danger. Neither is inherently good or bad; both should be intentional.

Managing covered calls when the stock rips higher

A covered call can become emotionally thorny when the stock rises quickly. You watch the shares rally and realize your upside is capped. This is where many traders commit the classic sin: paying too much to remove the cap *because they fell back in love with the stock mid-rally*.

The first question is not "How do I avoid being called away?" The first question is:

> Do I still want to sell my shares at the effective sale price I agreed to?

If yes, then do nothing. Being called away at a price you pre-approved is not a tragedy. It's a win you planned for.

If no, you have two clean choices:

- *Buy to close* the call (accepting the cost of removing the cap).
- *Roll up and out* (close the existing call and sell another call at a higher strike and later expiration, ideally for a net credit or a small, justifiable debit).

Rolling up and out can feel elegant: you keep the shares, raise the ceiling, and often collect additional premium. But be honest about what you're doing: you are rewriting the contract because the stock moved faster than you expected. If you keep rewriting every time a position becomes inconvenient, you can transform a disciplined strategy into a perpetual chase.

Managing covered calls when the stock falls

When the stock drops, the short call often loses value quickly. That can be a gift. If you sold a call for $2.00 and it's now worth $0.60 because the stock fell, you've already harvested a lot of income from that contract.

At that point, you have choices:

- Close the call early, lock in the profit, and sell a new call (possibly at a lower strike) if you still want to generate income.
- Leave it alone and let it expire, especially if it's far out of the money and not worth the attention.

The deeper question is about the stock itself. Covered calls do not solve "bad stock." If the decline reflects a fundamental deterioration, selling more calls can become a way of staying in denial while collecting small premiums against a large unrealized loss.

There's a difference between harvesting income on a temporarily depressed asset and clinging to a broken thesis because you can still sell calls. The market doesn't pay you extra for loyalty.

Managing cash-secured puts when the strike is tested

A put position becomes interesting when the stock approaches the strike. That's when your "paid bid" becomes a near-term purchase decision.

You have three honest paths:

Accept assignment. If you truly want the shares at the strike, the cleanest "management" may be to do nothing and let the contract play out. Assignment is not a malfunction; it is the trade's natural conclusion.

Close the put. If your view changed or you no longer want the shares, buy back the put and exit. This is the simplest way to stop the obligation.

Roll the put (often down and out). If you still like the underlying but want a lower entry price or more time, you can roll: buy back the current put and sell a later-dated put, possibly at a lower strike. This often reduces immediate assignment risk, but it also extends your commitment. You're trading calendar time for breathing room.

The trap is rolling as a reflex. If the stock is falling because the company is imploding, rolling down and out is not "defense." It's agreeing to buy later at a lower price while the elevator is still descending. Sometimes that's rational. Often it's a slow-motion way to turn "I might buy this stock" into "I am now married to this stock."

Managing credit spreads: defined risk, but precision required

Credit spreads come with built-in limits, which is exactly why people overtrade them. They feel safe, so traders stack them, and then discover correlation—the fact that many positions lose together in a real market selloff.

When a spread is threatened (the stock approaches or breaches the short strike), management options exist, but they are more technical because you have two legs:

Close the spread. Often the cleanest move. You take the loss (or small loss), free up margin, and live to sell premium another day.

Roll the spread out. Close the current spread and open a later-dated spread, sometimes adjusting strikes. This can collect additional credit and buy time, but it also extends exposure. Rolling spreads is not a magic eraser; it is a trade that can compound mistakes if you're rolling because you can't accept being wrong.

Reduce risk by changing the structure. In some cases traders will add a wing, convert to an iron condor, or otherwise reshape the position. These maneuvers can be valid, but they have a hidden cost: complexity. Complexity is not free. It demands attention and execution skill. If complexity exceeds your comfort, the “defense” can become the most dangerous part of the trade.

A simple rule keeps you sane: if you cannot explain your adjusted position in one or two sentences, you probably shouldn’t have it on.

Early assignment: the surprise you prevent by acting before it's ``urgent''

Short options can be assigned before expiration. That sentence is both obvious and routinely ignored.

Early assignment is less about random bad luck and more about predictable incentives. For covered calls, dividend timing can increase early assignment risk when the call is in the money and has little time value left. For puts, early assignment can happen too, often when the option is deep in the money and time value is small.

You don’t need to obsess, but you do need to monitor when the risk is logically higher:

- Short calls that are deep in the money near ex-dividend.
- Short options with very little time value remaining.
- Positions near expiration where the underlying is hovering near the strike.

This is why “I’ll just wait until expiration” is not a management plan. It’s a hope dressed as a plan.

A practical ``defense ladder": decide in advance how you respond

When the market moves against you, decisions feel urgent. Urgency makes people pay too much (to buy back options), roll too far (to avoid admitting loss), or double down (to get premium back). A defense ladder removes some of that heat by pre-committing to actions at certain points.

A workable ladder can be as simple as:

- If profit reaches a target (often around 50%), close and redeploy.
- If the short strike is tested, decide: accept assignment/call-away, roll, or close.
- If the underlying thesis changes, close—no negotiation.

Notice how little of that depends on prediction. It depends on behavior.

The emotional trap: ``rolling for a credit" can still be losing

People love the phrase "I rolled for a credit." It sounds like you got paid to fix the problem.

Sometimes you did. Sometimes you didn't.

Rolling for a credit can still be a losing sequence if:

- the credit is small relative to the risk you extended,
- you lowered your strike repeatedly (on puts) while the stock kept falling,

- or you capped your upside repeatedly (on calls) while the stock trended up.

Credits are not trophies. They are prices. The only question is whether the price is adequate for the obligation you're extending.

A disciplined trader can roll for debits when it's rational (paying to reduce risk), and can refuse to roll for credits when it's irrational (getting paid to remain in a bad deal). The market will happily pay you to do unwise things. That's what markets do.

Execution matters: bad fills are silent losses

Option management is often discussed like a chess game, but it's also a logistics job. Bid/ask spreads, liquidity, and order types matter. A sloppy close or roll on a wide market can donate a meaningful portion of your edge.

Two practical habits reduce friction:

- Use limit orders rather than market orders, especially on multi-leg spreads.
- Trade liquid strikes/expirations where bid/ask spreads are tight enough that you're not paying a hidden tax every time you touch the position.

It's not glamorous. It's the difference between "strategy works in theory" and "strategy pays in real life."

Actionable takeaway: write a one-page management policy and follow it for 10 trades

Create a simple management policy for your income trades and commit to following it for your next ten positions. Keep it short enough that you'll actually use it.

Include:

- Your profit-taking rule (e.g., close at 50% max profit, or another threshold you can live with).
- Your "tested strike" rule (accept assignment, roll, or close—decide which conditions trigger each).
- Your "thesis broken" rule (what counts as a fundamental change that forces an exit).
- A reminder to check for elevated early assignment risk (deep ITM calls near ex-dividend; near-expiration pin risk).

Then do the unglamorous part: follow it. Consistency is the whole point of an income factory. Rolling and defense aren't about outsmarting the market; they're about outlasting your own worst impulses while the market does what it does.

Chapter 6

Constructing Your Income Factory

You wouldn't build a house by randomly tossing bricks into a pile and hoping they form a wall. Similarly, picking a few high-yield stocks and selling a random put option doesn't constitute a strategy-it constitutes a gamble. In this chapter, we transition from understanding individual tools to drafting the blueprints for your entire factory. We will design a system that manages risk through structural rules, keeps the cash 'machinery' oiled, and runs smoothly whether the market is booming or busting.

6.1 Portfolio Design and Diversification

There's a moment every income investor hits—usually right after they've fallen in love with a *perfect* stock—when the market reminds them who's in charge. The company still looks great. The dividend still lands like clockwork. And yet the share price starts doing that slow, ugly staircase down. That's when the uncomfortable question shows up: *Was I building an income plan, or was I just collecting "good ideas"?*

Stock picking is a little like choosing the best bricks at the hardware store. Portfolio design is deciding whether you're building a cottage, a warehouse, or a bridge that has to hold real weight in real weather. An Income Factory isn't a list of tickers. It's an allocation—a deliberate blend of income *engines* that don't all stall at the same time.

The core idea is simple and surprisingly rare in practice: diversify not only *what* you own, but *how* you get paid.

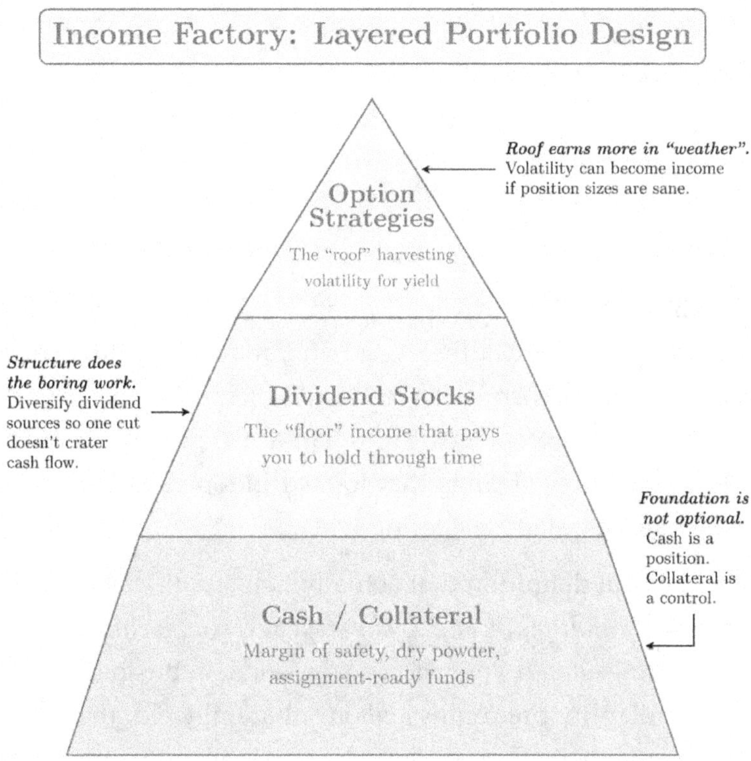

A layered approach solves a problem most income portfolios don't admit they have: they're usually built on *one* economic bet. Even if you hold twenty dividend stocks, you may still be betting on the same forces—consumer demand stays steady, credit stays available, input costs behave, management keeps its promises. When that bet goes sour, the portfolio doesn't just wobble; it starts marching in one direction, together.

Layering is a way to break that synchronized failure pattern. Cash behaves differently than stocks. Dividends behave differently than option premiums. And option premiums behave differently across volatility regimes—the market's "weather," where calm and storm are not moral judgments, just conditions you must plan for.

Diversification means ``different failure modes,'' not ``more tickers''

Most investors hear "diversification" and picture a fruit salad of holdings. Tech, healthcare, a bank, maybe a utility—done. The problem is correlation, the invisible string that ties different names into one giant trade. In a real drawdown, correlations can surge. Things that looked independent suddenly move like they share a steering wheel.

The practical definition that actually helps: *a diversified income portfolio has multiple income sources that fail for different reasons, on different schedules.* If your income depends on business profits *and* on volatility premiums *and* on collateral yield, the market must break you in three different ways at the same time to shut the factory down.

That's why "strategy diversification" is the Income Factory's version of sector diversification. You're not only mixing industries; you're mixing *paycheck types*.

The Dividend Floor: boring on purpose

Dividends are not magic. They're a distribution of corporate cash flows. But they have a superpower: they pay you for *waiting*. In markets where price action is irritating or slow, dividends can be the difference between a portfolio that feels alive and one that feels like dead capital.

Think of the dividend layer as a floor, not a rocket. Floors are supposed to be flat. They're supposed to hold furniture. If you try to turn the floor into a trampoline—chasing the highest yield you can find—you often get what looks like income but behaves like a trap: a shaky business paying an unsustainable dividend while the stock price quietly bleeds out the back door.

A stable dividend layer tends to have a few recognizable traits:

- The dividend is supported by a business that can endure a recession without existential drama.
- Management has a culture of paying shareholders *and* reinvesting enough to avoid slow decay.
- The balance sheet is not a knife-edge where higher rates or a credit scare forces a dividend cut.

Notice what's missing: *the highest current yield.* High yield can be fine, but it's not a virtue by itself. For a dividend floor, the job is reliability, not seduction.

One way to make that reliability real is to diversify the *source* of dividends the way you'd diversify your own paycheck. If all dividends ultimately depend on the same customer (say, the U.S. consumer), the floor isn't as sturdy as it looks. A blend across business models—staples, healthcare, infrastructure-like utilities, diversified financials, and selectively chosen real estate—creates a sturdier cash-flow base. Not because any one sector is "safe," but because their pain points are different.

The Option Overlay: turning ``market weather" into revenue

Options are where many investors either get reckless or get religious. Reckless: treating premium like free money. Religious: treating options as inherently sinful. Both miss the point. Options are tools that let you trade *shapes* of outcomes.

Selling covered calls and cash-secured puts is best understood as running a small insurance business. You collect premium for taking on specific risks:

- A covered call pays you now, in exchange for giving up some upside later (because you may have to sell at the strike price).
- A cash-secured put pays you now, in exchange for agreeing to buy shares at the strike price if assigned.

The overlay works because volatility—the market's habit of overpaying for certainty—often carries a risk premium. Many investors want protection *immediately*, especially when fear is in the air, and they'll pay up for it. When you're the seller, that fear can become your paycheck.

But here's the crucial design detail: the option overlay belongs on top of the dividend layer, not instead of it. Why? Because option income can be *lumpy*. It's sensitive to price swings, timing, and the simple reality that sometimes the market runs hard in one direction and your options position is the speed limit sign you installed on your own portfolio.

A well-designed overlay doesn't aim to maximize premium. It aims to *stabilize cash flow across different weather*:

- In choppy, sideways markets, covered calls can shine because you repeatedly sell time value while the stock goes nowhere.
- In nervous markets with elevated volatility, cash-secured puts can pay unusually well *if* you're comfortable owning the underlying at a discount and have the collateral ready.
- In strong bull markets, the overlay may lag because calls cap upside; that's not failure, that's the cost of getting paid for selling upside.

This is why it's called an overlay. The portfolio should still work if the overlay is turned down, paused, or temporarily unattractive. Your dividend floor and cash foundation keep the lights on.

Cash isn't lazy; it's structural

Traditional investing culture treats cash as a guilty pleasure. If cash is sitting there, you're "missing out." Income Factory thinking treats cash as a *load-bearing wall*. It does three jobs at once:

1. It's collateral for cash-secured puts (and a buffer against forced decisions).

2. It's dry powder for rebalancing when markets hand you discounts.
3. It's a shock absorber that reduces the chance you'll sell good assets at bad times.

The emotional value of that last point is underrated. A portfolio that is always 100% invested is like a household with zero emergency fund: it can work for years, then one surprise expense turns into a crisis. Cash reduces the number of scenarios where you have to do something you'll regret.

From a design standpoint, the cash layer is also what makes option selling honest. A cash-secured put is only truly "secured" if the cash is actually there, reserved, and not quietly promised to three other trades. If the cash isn't real, you're not selling puts; you're borrowing leverage from your future self.

How the layers interact: a practical mental model

A portfolio built from layers behaves differently than a portfolio built from picks.

Imagine three common market regimes:

1) Calm market, slow grind upward. Dividends keep paying. Volatility tends to be lower. Covered calls may generate smaller premiums, and they may occasionally be tested as prices rise. In this regime, the overlay is not the star of the show; it's a steady side business. The key is not to sell calls so aggressively that you spend the whole year watching your best holdings get called away just as they begin to run.

2) Sideways, choppy market. This is the environment many investors hate and many income strategies enjoy. Prices whip around; headlines rotate; nothing "happens" for months. Dividends do their job. Covered calls can be surprisingly powerful because you can sell premium repeatedly while the underlying fails to trend. Your portfolio can feel like it has traction even when the index is stuck in mud.

3) Sharp selloff, fear-driven volatility. This is where design is exposed. Dividend stocks drop (sometimes brutally), and weak dividend stories get punished twice: price down *and* dividend credibility questioned. But volatility usually rises, which can make put premiums richer and call premiums fatter.

It's tempting to see that rich premium as opportunity and sell as much as possible. That's how people turn a bad week into a bad year. In a true selloff, the option overlay must be governed by the foundation: you only sell the amount of puts your cash can actually support, and you choose underlyings you'd be willing to own through ugliness. The cash layer is what keeps you from panicking, and it's also what keeps you from overcommitting when the market is shouting at you.

The point isn't that the layers eliminate drawdowns. The point is that they give you *more levers*. When one lever stops helping, another may start.

Strategy diversification is also psychological diversification

A quiet benefit of blending dividends with options is that it changes how you experience the market.

If your only income source is dividends, you may obsess over dividend cuts and payout ratios. Every earnings report feels like an audition. If your only income source is options, you may watch the screen like a weather radar, unable to relax because you're always one gap-down away from a new problem.

Blend them and something weird happens: you start caring less about being right. Dividends reward patience. Option premium rewards process. Cash rewards humility. Together, they push you toward a professional mindset: focus on repeatable actions, not heroic predictions.

That psychological diversification matters because most portfolio blow-ups are not math failures. They're behavior failures. People abandon good systems at the exact moment those systems are designed to pay off. A layered Income Factory is built to be easier to stick with.

A workable blueprint: building the allocation without pretending you can forecast

Asset allocation is where investors often ask for a precise recipe. Percentages. Exact numbers. A guarantee that if they do it "right," they won't feel pain.

No such guarantee exists, but a sensible blueprint does.

Start by assigning each layer a job:

- *Cash/Collateral:* keep you solvent, flexible, and able to take assignments without panic.
- *Dividend Stocks:* produce baseline cash flow and long-term inflation fighting.

- *Option Overlay:* harvest premium when it's attractive, enhance yield, and monetize volatility without becoming the whole portfolio.

Then decide what you want the factory to optimize:

- If you need steady, sleep-at-night income, you bias toward a thicker dividend floor and more cash.
- If you want higher cash flow and can manage positions actively, you allocate more to the option overlay—but you compensate with stricter rules on collateral and notional exposure.

Here's the design trick that keeps this from becoming guesswork: treat the option overlay like a *dial*, not a switch. You don't have to be 0% or 100% "in options." You can run a light overlay in calm markets and a more selective, collateral-heavy posture when volatility is high. The allocation can be stable even when the activity level changes.

Case study: two investors, same stocks, different architecture

Picture two investors who both own a basket of dividend stocks.

Investor A puts 95% in those stocks and keeps 5% in cash. Their income is almost entirely dividends.

Investor B holds the same dividend basket, but keeps a meaningful cash sleeve earmarked for cash-secured puts, and runs covered calls on a portion of the equity positions.

Now the market goes sideways for a year. Not a crash, not a boom—just chop.

Investor A collects dividends and feels increasingly annoyed. The portfolio statement looks like a treadmill. They start tinkering: swapping out "boring" dividend names for higher yield, then higher yield for higher growth, then back again. The danger isn't the returns; it's the restless behavior that returns provoke.

Investor B collects dividends too, but the overlay creates additional cash flow *because the market is choppy*. Selling covered calls against stable holdings generates recurring premium. Cash-secured puts may be sold selectively on names they're happy to own at lower prices. Investor B feels less pressure to find a new "great idea" because the architecture itself is productive.

Same stocks. Different experience. Different outcome potential. That's portfolio design.

The hidden diversification problem: your income can be diversified while your risk is concentrated

A portfolio can produce income from many places and still be concentrated in one risk factor.

Examples:

- Owning ten high-yield stocks that all depend on cheap credit is credit-risk concentration.
- Selling puts across many tickers that all drop together in a risk-off shock is correlation concentration.
- Running covered calls on the same high-beta sector is volatility concentration.

The fix is not perfection. The fix is to regularly ask a blunt question: *If one macro lever moves against me—rates jump, oil spikes, consumers retrench, volatility explodes—how many of my "diversified" positions are really the same trade?*

This is also where index-based thinking can help. Many investors feel they must choose between "individual stocks" and "ETFs." For the Income Factory, that's a false choice. Broad funds can provide structural diversification for the dividend floor, while options activity can be focused on highly liquid underlyings where spreads are tighter and management is easier. Architecture first; instruments second.

Rebalancing: the unglamorous habit that keeps the factory from leaning

Diversification is not a one-time decision because markets are not static. Even if you start perfectly balanced, drift happens:

- A winner grows and quietly becomes a concentration.
- A sector runs hot and your portfolio becomes a bet you didn't mean to place.
- Option premium tempts you into doing more of what recently worked.

Rebalancing is simply restoring your intended weights. It feels counterintuitive because it forces you to do the opposite of momentum: trim what has grown and add to what has lagged *when the thesis still holds*. It's not a performance trick. It's a risk-control practice.

For an Income Factory, rebalancing has an extra twist: you can often rebalance with *cash flows* instead of selling.

- Dividends can be directed toward underweight areas.
- Option premium can be treated as replenishing the cash foundation first, then deployed.
- New contributions can be used to correct drift without creating taxable sales.

That's architecture doing what architecture does: it turns maintenance into routine, not drama.

Actionable takeaway: write a one-page ``Income Factory allocation charter"

Do this in plain language, on one page, and keep it where you place trades. Your future self will need it on a bad market day.

Include:

- *Your three layers and their target ranges* (not a single number): Cash/Collateral, Dividend Stocks, Option Overlay.
- *The job of each layer* in one sentence.
- *A dial rule for the overlay:* what conditions make you run it lighter, and what conditions make you run it only if collateral is abundant.
- *A rebalancing trigger:* a simple rule like "if any layer drifts more than X percentage points from its range, I rebalance using cash flows first."

If you can't explain your allocation charter without referencing a specific ticker, you don't have a design yet—you have a collection. The goal is a factory that runs because it was built to run, not because the market happens to be sunny.

6.2 Sizing and Concentration Limits

The fastest way to blow up an income portfolio is not picking a bad stock. It's picking a *normal* stock and making it a *life-or-death* position without realizing it.

Concentration risk almost never arrives wearing a villain costume. It shows up as a success story: one holding runs, you don't sell, and now it's 18% of your account. Or it shows up as a "safe" income idea that everybody owns, so you size it up because it feels familiar. Or it sneaks in through options, where a handful of trades can quietly promise your future self a mountain of stock you can't comfortably carry.

Sizing rules are the factory's circuit breakers. They don't make you smarter; they make you harder to kill.

Why income investors are especially vulnerable to over-sizing

Growth investors often worry about one thing: pricc. Income investors worry about two: price *and* cash flow. That second worry creates a dangerous itch—the urge to fix a cash-flow shortfall by "just" making the next trade bigger.

If you've ever caught yourself thinking, *If I sell twice as many puts, I'll get twice the premium*, you've already met the trap. Premium scales linearly; risk does not. Risk has corners. It has jump gaps. It has the nasty habit of clustering when you're least prepared to handle it.

The market doesn't need you to be wrong ten times. It only needs you to be overexposed *once*.

The prime directive: any single position should be allowed to fail

A real sizing framework starts with a slightly grim but incredibly freeing thought experiment:

> If this position dropped 50% (or worse), would it be an inconvenience—or would it change my life?

You're not predicting a 50% drop. You're asking whether your portfolio can survive a bad headline, a broken business model, an accounting scandal, a regulatory hit, or a recession that treats your favorite sector like a punching bag.

When the answer is "it would change my life," the fix is not better research. The fix is smaller size.

The 5% rule (and why it feels too conservative right up until it doesn't)

A blunt, effective default for individual stocks is a maximum of about 5% of portfolio value per position at cost (some investors use market value; either is fine as long as you're consistent). Five percent is not sacred. It's just a number with a nice property: it makes disasters survivable.

If a 5% position gets cut in half, the portfolio takes about a 2.5% hit. That stings. It does not end you. If a 20% position gets cut in half, that's a 10% portfolio drawdown from one mistake. That's how investors end up in the psychological danger zone where they abandon plans, revenge-trade, or "take a break" for three years and miss the recovery.

Here's the uncomfortable truth: if you need a single position to be bigger than 5% to make your income plan work, the plan is fragile. You're depending on concentration for cash flow.

A more flexible way to use the 5% rule is as a *default ceiling*, not a law of nature:

- Broad, diversified funds can reasonably be larger than 5% because they already contain many businesses.
- A single company with binary risk (litigation, regulatory exposure, commodity dependence, high leverage) often deserves less than 5%.

This is not about bravery. It's about acknowledging that some businesses come with hidden trapdoors.

The ``double-counting" problem: sector exposure that pretends to be diversified

Owning twelve dividend stocks can still be one bet if they all lean on the same pillar. Investors learn this the hard way when a whole sector gets repriced overnight. It feels unfair until you notice the common thread: the holdings were diversified by *name*, not by *driver*.

A simple, usable sector limit is something like 20%–25% in any one sector for an individual-stock portfolio (many investors go lower). Again, the exact number matters less than having a number.

What makes this tricky in an Income Factory is that your exposures may be disguised:

- "Income" portfolios often drift toward rate-sensitive sectors (utilities, REITs, telecom), so a change in interest-rate expectations can hit multiple holdings at once.
- High-yield screens often concentrate you in leveraged balance sheets, which tend to share the same failure mode when credit tightens.

If you want a quick gut check, ask: *What headline would hurt most of my holdings at the same time?* If the answer comes quickly, you're more concentrated than you think.

Options sizing: the part most people get wrong because brokers make it look easy

A broker can show you a neat little line that says "credit received: $237." Your brain registers: *small trade, small risk.*

But selling options is like signing a contract that only becomes important when things go sideways. The risk isn't the premium you collect. The risk is the obligation you've agreed to.

Two definitions you must treat like shop safety goggles:

Notional exposure.

Notional exposure is the dollar value of the shares you're effectively on the hook for. For a short put, it's simple:

$$\text{Notional Exposure} = \text{Strike Price} \times 100 \times \text{Number of Contracts}.$$

Sell one put with a $50 strike? You've promised you can buy $5,000 of stock. Sell five contracts? You've promised $25,000.

This is where option sellers accidentally concentrate. They look at premium and forget that the contract scales in blocks of 100 shares.

Assignment risk.

Assignment is when you're obligated to buy (puts) or sell (calls) shares because the option buyer exercised. For U.S. equity options (typically American-style), assignment can happen before expiration. That means your sizing plan must survive a scenario where you wake up to shares in your account *earlier than you expected.*

If that possibility makes you queasy, it's not a sign that options are "dangerous." It's a sign you're about to size them in a way that makes you dependent on perfect timing.

A practical rule: collateral first, premium second

A cash-secured put is only as conservative as the cash behind it.

The most factory-friendly approach is to treat each short put as a reserved parking spot for capital. The cash is not "available." It's spoken for. When you place the trade, you should already be comfortable with this sentence:

> If assigned, I will own 100 shares at the strike price, and my portfolio will still be within its sizing limits.

That sentence forces three disciplines at once:

- You're choosing underlyings you'd be willing to own, not just symbols with fat premiums.
- You're limiting contract count to what your cash and concentration rules can support.
- You're treating assignment as a normal outcome, not a catastrophe.

Notice what this removes: the fantasy that you'll always "roll" or "close" in time. Rolling can be a useful tool. It's not a safety plan. In a fast market, liquidity widens, spreads bite, and the roll you imagined can become expensive or unavailable at the exact moment you want it most.

Position size rules that actually work (because they're boring)

Here are sizing constraints that are simple enough to follow when you're tired, busy, and tempted.

Rule 1: cap any single-stock exposure after assignment.

Before you sell a put, calculate the *post-assignment* position size:

$$\text{Post-Assignment Size} = \frac{\text{Strike} \times 100}{\text{Portfolio Value}}.$$

If assignment would create a 9% position and your cap is 5%, the trade is oversized even if your broker calls it "cash-secured."

This one calculation prevents a huge category of option-selling mistakes.

Rule 2: cap total assigned-equity exposure from short puts.

You can also cap how much of your portfolio could turn into stock if everything went against you at once. Call this your *put-assignment budget*. Conservative investors might keep it at, say, 10%–25% of the portfolio; more aggressive investors might go higher, but only with eyes open and with a clear reason.

Why it matters: when markets fall hard, *multiple* puts can land simultaneously. The danger isn't one assignment. It's five assignments plus a shrinking portfolio value plus the sudden realization that your "cash" was already committed.

Rule 3: diversify expirations the way you diversify stocks.

This sounds small, but it's a major stress reducer. If all your options expire the same week, you've created a single calendar cliff. Spreading expirations means you're not forced to make all your decisions under the same market mood.

It also helps you avoid the classic income-investor mistake: harvesting premium in calm weeks and then finding yourself overloaded right when volatility spikes.

Covered calls: income with an opportunity-cost price tag

Covered calls are often described as "conservative" because you own the shares. That's partly true. Your downside is the stock's downside (minus the premium you collected). The bigger risk is different: you can accidentally sell away your best outcomes.

A covered call is a contract that says: *I'm willing to sell at this price.* If you aren't truly willing, the call is not covered emotionally, even if it's covered mechanically.

Sizing for covered calls has two dimensions:

- *Share position size:* still subject to your single-stock cap. The call doesn't reduce the risk of owning too much of one company.
- *Call coverage ratio:* what portion of your shares you're willing to have called away. Many investors feel more stable starting with partial coverage (for example, selling calls against 25%–50% of a position) until they learn how it feels to watch a stock run through their strike.

There is a very real emotional cost to being "right" about a company and still underperforming because you sold calls too tightly. The fix is not avoiding calls. The fix is treating call selling as a yield-enhancement tool with a budget for regret.

Kelly sizing: brilliant math, dangerous vibes

At some point, you may encounter Kelly criterion sizing—a formula that, in theory, maximizes long-run growth by scaling position size to edge and odds. It's elegant. It's also the kind of elegant that can get people hurt when it meets real markets.

Two issues make Kelly-style sizing risky for most individual investors:

- *You don't know your true edge.* Your "win rate" and payoff distribution are not stable, especially in options where volatility regimes change.
- *The drawdowns can be psychologically unlivable.* Even when Kelly is mathematically optimal, it can be emotionally catastrophic, which means you abandon it at the worst time.

If you like the spirit of Kelly (sizing based on risk, not vibes), steal the part that matters: *when uncertainty is high, size down.* In option selling, uncertainty often shows up as elevated volatility and jumpy price action. That's not a cue to swing harder because premiums are juicy; it's often a cue to demand a bigger margin of safety.

A ``factory-safe'' sizing worksheet (do it before you trade, not after)

Before entering any new position, answer these questions in writing. Not as a ritual. As a way of slowing your hands down enough to keep your brain involved.

1. *If this stock dropped 50%, what would happen to my portfolio?*
2. *If I were assigned on every short put I have open, what percentage of my portfolio would turn into stock?*
3. *After that assignment, would any single name exceed my cap? Would any sector exceed its cap?*
4. *If shares are called away, will I feel relieved (planned exit) or angry (unplanned loss of upside)?*

The questions are simple. Their power is that they force you to imagine the stressful version of reality *before* it happens.

The real goal: turning disasters into annoyances

Sizing rules aren't about eliminating losses. Losses are the cost of admission. The goal is to keep losses from changing your identity as an investor.

When position sizes are sane, a dividend cut is a headache, not a personal betrayal. An assignment is a workflow event, not a crisis. A sector slump is an opportunity to rebalance, not a reason to abandon the plan.

You can't control the market's mood. You can control the size of your promises.

Actionable takeaway: adopt three hard caps and one ``pre-trade" calculation

Write these down and treat them like safety rails, not suggestions:

- A maximum size for any single stock position (start with 5%).
- A maximum exposure to any single sector (pick a number and make it real).
- A maximum total notional exposure from short puts that could be assigned (your put-assignment budget).

Then, for every short put you consider selling, do one calculation before you click "sell":

$$\textit{Post-assignment position size} = \frac{\text{Strike} \times 100 \times \text{Contracts}}{\text{Portfolio Value}}.$$

If that number violates your cap, the trade is not "a little too big." It's structurally unsafe. Reduce contracts, choose a cheaper underlying, or don't take the trade. The factory stays open by refusing to sign contracts that could shut it down.

6.3 Cash Management and Deployment

Cash has a public reputation problem. In investing circles, admitting you hold cash can feel like admitting you showed up to a marathon with a folding chair. People whisper about "cash drag" and missed rallies, as if every dollar not fully deployed is a moral failure.

That attitude is how otherwise careful investors end up doing something truly expensive: they turn their portfolio into a machine that must be *perfectly timed* to keep paying the bills. Cash is the antidote to perfect timing. It's the part of the Income Factory that doesn't care what the market thinks today.

The trick is to stop thinking of cash as an empty parking space and start thinking of it as *infrastructure*. In an options-and-dividends portfolio, cash has jobs. Multiple jobs. And when cash has jobs, you manage it like a manager—with rules, schedules, and a clear definition of what "available" actually means.

Three kinds of cash (and why mixing them causes trouble)

In an Income Factory, the word "cash" covers three very different things that only look similar on a brokerage statement.

1) Spendable cash. This is the money you can use for life without triggering a forced decision elsewhere. Mortgage, groceries, tuition, property taxes, the annoying surprise expense that arrives precisely when the market is having a tantrum. Spendable cash is your *operations budget*.

2) Collateral cash. This is cash reserved to keep your promises. If you sell a cash-secured put, the cash is not idle—it's the collateral that makes the trade honest. The premium you collected is the visible part; the collateral is the load-bearing part.

3) Strategic cash. This is the optionality bucket. Dry powder for rebalancing, for buying when prices are down, for increasing option activity when premiums are attractive. Strategic cash is not earmarked for next month's bills, and it's not already pledged to an options contract.

If you treat all three as one blob, you will eventually steal from one job to pay another. That's when the factory starts running on stress.

Collateral is a holding tank, not a cost

When you sell a cash-secured put, the broker typically earmarks cash (or cash-like buying power) against your obligation to buy shares at the strike price. Psychologically, many investors experience that earmarked cash as "dead." They stare at it and think, *I'm not earning anything on that money, so I should use it.*

That thought is exactly backwards. Collateral is doing work even when it's sitting still:

- It keeps you from overcommitting notional exposure.
- It lets you accept assignment calmly instead of scrambling.
- It prevents the nightmare scenario where a down market forces you to sell something else at the worst possible moment just to cover an obligation.

In other words, collateral converts a potentially leveraged activity into a controlled, cash-governed one. The "return" on collateral is not only yield. It's survivability.

And yes, collateral can earn yield too, depending on where you hold it and what your broker pays on cash balances. That yield is a nice tailwind. It is not the core reason collateral exists.

The cash ladder: deploying money without needing a crystal ball

Most investors deploy cash like they rip off a bandage: all at once, hoping they chose the right moment. That's how cash becomes a source of regret. If the market rises, you feel stupid for holding it. If the market falls, you feel stupid for deploying it.

A cash ladder is a simple way to turn deployment into a schedule instead of a prediction. The idea: split strategic cash into tranches and pre-commit to when (or under what conditions) each tranche gets deployed.

Here are two ladder styles that work well in an income portfolio:

Time ladder. Divide strategic cash into, say, six to twelve equal pieces and deploy one piece per month (or per quarter). This is not exciting. That's the point. You're buying relief from timing anxiety.

Price/volatility ladder. Deploy tranches when the market gives you a reason, such as:

- a broad index down a certain percentage from its recent high,
- or volatility noticeably elevated (which often improves option premiums),
- or a watchlist stock reaching a pre-chosen valuation/price zone.

The rule matters more than the trigger. You want something objective enough that Future You can't negotiate with it when fear is loud.

In a factory mindset, ladders are how you keep cash from becoming either a permanent couch potato or a panicked hero.

Cash-secured puts as a deployment tool (with a speed limit)

The cash-secured put is often pitched as "getting paid to buy stocks." That's catchy and mostly true, but it hides the operational detail that matters: the put is a *conditional purchase order*.

When you sell a put, you are saying:

> I will buy 100 shares at this price, and I'm being paid for standing ready.

That makes cash-secured puts a natural bridge between cash and equity exposure:

- If you *aren't* assigned, you've rented out your patience for premium.
- If you *are* assigned, you've deployed cash into shares at an effective price reduced by the premium collected.

But the speed limit is crucial: you can't sell more puts than your collateral cash can cover *and* still call it an Income Factory. That becomes an assignment trap, where multiple puts land at once and your "cash" turns out to be a pile of promises.

A clean way to keep the speed limit real is to allocate a specific slice of your portfolio to a "put inventory" program. Think of it as a production line with a maximum capacity. If the capacity is full, you don't jam more raw material into the machine. You wait.

The withdrawal problem nobody wants to solve: selling into a down market

Even investors who love dividends and options often fund their real-life spending the old-fashioned way: when they need cash, they sell something.

That works beautifully until the market drops 20% and your brain has to do this math in real time:

> If I sell now, I'm locking in losses *and* shrinking the base that produces my future income.

This is sequence risk: the risk that poor returns early in your withdrawal period do outsized damage, because withdrawals force you to sell more shares at depressed prices. It's not theoretical. It's the difference between a plan that looks fine on paper and a plan that feels cruel in practice.

Cash management is how you buy your way out of that trap.

The spending buffer: the most underrated ``investment" you can own

A spending buffer is a dedicated pool of spendable cash (or very cash-like instruments) meant to cover a set period of living expenses. Many investors choose something like 6–24 months, depending on income stability, risk tolerance, and whether portfolio withdrawals are a primary funding source.

This is not about maximizing returns. It's about maximizing *choices*.

When the market is down, the buffer lets you:

- keep paying yourself without selling depressed assets,
- keep writing options from a position of calm (instead of desperation),
- and let dividends do their slow, reliable work without demanding they solve every short-term cash need.

A buffer also improves behavior. When people don't *need* the market to cooperate this month, they make better decisions. They size better. They chase less. They roll less impulsively. They stop trying to wrestle cash flow out of the market with their bare hands.

Where to hold cash: the boring details that matter in bad weeks

Cash can live in several places: a bank, a brokerage sweep, a money market fund, Treasury bills, short-term Treasury ETFs, and so on. The right choice depends on your goals and your account type, but one principle survives every environment:

> If you might need the cash on short notice, you should understand how it becomes cash again.

In calm markets, liquidity feels infinite. In stress, it becomes a feature you suddenly care about intensely. Some cash instruments have settlement timing, trading hours, or redemption mechanics that don't matter until the exact moment they matter most.

The practical advice is not "avoid everything complicated." It's this: match the cash vehicle to the cash job.

- Spendable cash wants simplicity and immediate availability.
- Collateral cash wants stability and operational compatibility with your broker's option mechanics.
- Strategic cash can tolerate a little more structure if it improves yield, as long as liquidity remains sufficient for your plan.

If you can't explain to yourself how quickly you can convert the instrument into spendable dollars, it's not a cash tool; it's an investment you happen to call cash.

A cash workflow that keeps the factory humming

Once you separate cash by job, you can build a workflow that feels almost mechanical. Here's a practical model many income investors find calming because it removes constant decision-making:

Step 1: Income arrives into a cash hub. Dividends and option premium land in one place (a designated sweep or settlement fund). Don't immediately reinvest them by reflex. Let them accumulate until the next scheduled decision point.

Step 2: Refill the spending buffer first (if it's below target). If your buffer is designed to cover, say, 12 months of expenses, treat that target like oil in an engine. When it runs low, you top it up before you think about new trades.

This one habit prevents the classic error: chasing yield because you're underfunded for life.

Step 3: Refill collateral capacity (if you run a put program). If your put program has a planned capacity—for example, you want to keep enough collateral to sell a certain number of contracts within your sizing rules—then premiums can rebuild that capacity over time, especially after assignments.

Step 4: Deploy the remainder via the ladder. Only after life-cash and promises-cash are healthy do you deploy toward new equity positions, additional covered calls, or new put sales.

This order is not about being timid. It's about ensuring the factory can pay the electric bill before it buys a new machine.

Cash and options: the subtle benefit of being able to wait

Waiting is a financial edge most people refuse to admit exists, because it feels passive. In options, the ability to wait is often the difference between selling premium at a fair price and selling premium at a desperate price.

When you have strategic cash and a spending buffer, you can be picky:

- You don't have to sell puts on junky underlyings just because the premiums look juicy.
- You don't have to sell covered calls so tight that you strangle your upside, simply because you need income *right now*.
- You can close or roll positions when it's sensible, not when your checking account is pressuring you.

Cash turns "I hope this works" into "I can afford to be selective."

That selectivity has compounding effects. It improves your average trade quality, your stress level, and your ability to stick to position-sizing rules when markets get loud.

A case study: the difference between cash as slack and cash as strategy

Consider two investors who both run dividend stocks with occasional cash-secured puts.

Investor A keeps very little cash because they hate "missing out." When the market drops sharply, they see their put premiums explode upward (tempting) and their portfolio value drop (threatening). They sell more puts to "make up" income, commit a large portion of remaining cash as collateral, and then get assigned on multiple positions as the market continues down. Now they're fully invested in a falling market with no flexibility. Any new opportunity is something they have to watch, helplessly, from the sidelines.

Investor B keeps a dedicated spending buffer and a planned collateral sleeve. In the same drop, they can sell *some* puts into higher premiums without violating their assignment budget. If assigned, they have a workflow: shares land, some become candidates for covered calls, and the spending buffer prevents forced selling. Investor B is not immune to drawdowns, but they maintain the ability to make choices. That's the whole game.

Same tools. Different cash governance. One is reactive; one is industrial.

Actionable takeaway: create a three-bucket cash policy and automate your decisions

Write a simple cash policy that fits on an index card:

- *Spendable cash target:* enough to cover ______ months of expenses.
- *Collateral cash target:* enough to support ______ cash-secured put assignments within your sizing limits.
- *Strategic cash ladder:* deploy ______% (or $______) on a schedule, plus an extra tranche if the market drops ______% or volatility spikes meaningfully.

Then schedule one recurring "cash meeting" with yourself (monthly works well). On that day, you move money according to the policy and place trades only if they fit the ladder and the collateral rules. The goal is to turn cash deployment from a mood into a system.

Cash isn't what you hold when you don't know what to do. Cash is what lets you keep doing the right things when the market gives you every reason not to.

6.4 Portfolio Archetypes

Two investors can hold the same stocks, sell the same options, collect the same dividends—and still have completely different outcomes. Not because one is smarter. Because one has a portfolio that matches their temperament, and the other is constantly at war with their own wiring.

That mismatch is the silent killer of income plans. The spreadsheet looks fine. The rules are sensible. Then real life shows up: a busy workweek, a surprise expense, a sleepless night during a market slide. The portfolio that required calm, attentive management suddenly belongs to someone who has neither calm nor attention to spare.

A good archetype is a portfolio you can run *as the person you actually are*. Not the person you become in a motivational burst on a Sunday afternoon.

What follows are three Income Factory "factory models." Each one is a coherent blend of dividends, cash governance, and options activity—but the emphasis shifts. The goal isn't to crown a winner. The goal is to pick a machine you can keep operating when the market is loud and your schedule is not negotiable.

Archetype 1: *The Fortress* (conservative, dividend-heavy, low maintenance)

The Fortress investor wants income to feel like gravity: always there, quietly doing its job. They're not trying to squeeze every last basis point from premium. They're trying to avoid the two most expensive emotions in investing: panic and impatience.

The Fortress is built around a thick dividend core and a deliberately meaningful cash reserve. Options are allowed, but only in a way that never threatens the foundation. Think of options here as a condiment, not the meal.

Who it fits.

- Investors funding real-life expenses from the portfolio (retirees, semi-retirees, or anyone with irregular income).
- People who don't want to monitor markets daily.
- Anyone who knows their greatest enemy is overtrading.

How it earns. The Fortress earns primarily from dividends and secondarily from *occasional* option premium. The big win is not maximum yield; it's consistency. The portfolio is designed to keep paying even when the market is doing its impression of a washing machine.

A typical Fortress cash-flow rhythm might look like this: dividends arrive, refill the spending buffer if needed, then the remainder gets reinvested into the dividend basket or used to rebalance. If volatility spikes and premiums are attractive, the Fortress may sell a small number of cash-secured puts on high-quality names it would happily own anyway—but only within a strict assignment budget.

What it gives up. The Fortress usually gives up the thrill of "making something happen" every week. It may underperform in roaring bull markets if it holds more cash and runs fewer aggressive overlays. That's not a flaw; it's the fee you pay for durability.

The hidden danger. Fortress investors often fall into a specific trap: treating dividend yield as a scoreboard. When the market is down, high yields can look like bargains. Sometimes they are. Sometimes they're a warning label. The Fortress stays a fortress only if it resists turning into a museum of distressed high-yield names.

A clean default blueprint. Many Fortress-style portfolios end up roughly in this neighborhood (adjust to your needs, not to a dogma):

- A substantial spendable-cash buffer (months of expenses, not days).
- A diversified dividend core (often best achieved with funds/ETFs if individual-stock diversification is hard).
- A small, rule-governed option sleeve focused on *cash-secured* puts and selective covered calls.

The Fortress feels slow until you live through a real drawdown and realize you never had to bargain with yourself.

Archetype 2: *The Harvester* (balanced, active covered calls)

The Harvester is the investor who wants their portfolio to feel like a working farm: reliable crops (dividends) plus seasonal harvesting (option premium). They're comfortable with routine management and they enjoy improving results through process rather than prediction.

The Harvester's signature move is covered calls: selling call options against stock positions to collect premium. Premium is the harvest. The stock position is the field. You're not trying to win the lottery; you're trying to monetize time and volatility.

Who it fits.

- Investors who can check positions weekly (not necessarily daily).
- People who enjoy systems and consistency more than "big ideas."
- Anyone who wants higher cash flow but still wants the psychological stability of owning quality equities.

How it earns. The Harvester earns from dividends *and* a steady cadence of call premium. The most satisfying Harvester experience tends to happen in markets that are sideways or moderately bullish. In those environments, covered calls can feel like getting paid for the market's indecision.

The Harvester's portfolio tends to have a well-defined "overwrite policy"—meaning how much of the stock portfolio will have calls sold against it. A common choice is partial coverage: not every share is overwritten, and not every holding is treated the same. Some positions are long-term compounders that you're reluctant to cap; others are mature, slower-moving names where call income is a natural fit.

The real cost: you are selling away upside. A covered call is not free money. It's a trade: premium now in exchange for some upside later. When the stock rallies hard, you can feel like you did something wrong—even though you did exactly what you agreed to do.

That emotional friction is why the Harvester needs a rule that protects against self-sabotage. One of the simplest is a "regret budget":

> I will sell calls at strikes where I would be genuinely willing to sell the shares, and I will avoid call-selling so tight that I'm constantly angry at my own portfolio.

This sounds soft. It's not. It's behavioral risk management.

Assignment: not a disaster, a workflow event. Calls can be assigned. Shares can be called away, potentially before expiration. A Harvester is prepared for this operationally and emotionally. If assignment occurs, the plan might be:

- accept the sale and redeploy into a similar holding,
- or wait for a pullback and re-enter,
- or rotate into a new candidate with better dividend quality or better option liquidity.

The point is not the exact tactic. The point is that the Harvester expects assignment as a normal outcome and has a decision tree ready before it happens.

The hidden danger. The Harvester's main risk is turning "covered calls" into "constantly selling calls no matter what." When volatility is low, premiums may not justify capping upside. When a stock is poised for a sharp move (earnings, major news, takeover chatter), call selling can become an unintentional bet against your own position.

The Harvester stays healthy by being selective: you harvest when there's something worth harvesting.

Archetype 3: *The Trader* (aggressive, high put-selling rotation)

The Trader is the investor who wants the Income Factory to feel like a production floor: frequent activity, measurable output, constant optimization. They are not necessarily reckless. But they accept that short option premium is a business with real risk, and they're willing to manage that risk actively.

The Trader's main tool is the cash-secured put program, often run in a rotation. The idea is to repeatedly sell puts on a set of liquid underlyings, collecting premium as "inventory income." If assigned, the Trader either holds the shares and sells covered calls (a wheel-style approach), or uses predefined rules to exit and reset.

Who it fits.

- Investors who can monitor positions frequently and respond quickly.
- People with a strong appetite for process and a low appetite for ambiguity.
- Anyone who enjoys options mechanics and understands that premium is compensation for risk, not a coupon.

How it earns. The Trader earns heavily from option premium and the yield on collateral cash. In many market environments, especially when volatility is healthy, this can produce very attractive cash flow. It can feel like the portfolio has a paycheck even when markets are messy.

But the Trader must internalize what they are *really* doing. Selling puts systematically is economically similar to selling insurance against market declines. It often wins in small increments and loses in ugly chunks. That's not pessimism; it's the shape of the payoff.

Regime dependence: the market changes the rules midgame. Option selling has seasons. Calm markets can starve premium. Volatile markets can offer rich premium *and* higher odds of assignment. Sharp crashes can overwhelm careless sizing. The Trader archetype works best when it is designed to *de-risk automatically* as conditions worsen.

A practical way to express that is a volatility-aware throttle:

- When volatility is low and premiums are thin, the Trader reduces activity rather than reaching for risk.
- When volatility rises, the Trader may sell puts with wider margins of safety (lower deltas, lower strikes) and fewer contracts.

The key is not the exact metric. The key is the habit: when the market pays you more to take risk, you do not necessarily take *more* risk. You can take *better-priced* risk at the same or lower size.

The hidden danger. The Trader's big enemy is not a single losing trade. It's accumulation of promises. You sell puts, then sell more, then roll, then sell again, and suddenly your account is a web of obligations that all point in the same direction: you are long equity risk whether you admit it or not.

This is why the Trader archetype lives or dies by strict notional exposure limits and a hard assignment budget. Without those, it's not a strategy. It's a slow-motion margin call waiting for a catalyst.

Choosing an archetype: the most important question is not ``what yields more?"

The best archetype is the one you'll still run when:

- the market is down 15% and headlines are spicy,
- your calendar is full,
- and your last three trades didn't work the way you expected.

That's when the portfolio meets the investor. A high-yield strategy that you abandon is a low-yield strategy in real life.

So choose based on constraints:

- *Time constraint:* How many minutes per week can you reliably give this?
- *Stress constraint:* How well do you sleep with open obligations?
- *Income constraint:* Do you need monthly cash flow, or can you tolerate lumpiness?
- *Behavior constraint:* Are you more likely to overtrade or to freeze?

These are not personality quiz questions. They are engineering inputs.

A simple way to blend archetypes without creating a Frankenstein portfolio

Many investors aren't purely one archetype. That's normal. The mistake is mixing styles in a way that multiplies complexity without adding resilience.

A clean approach is to choose one archetype as the *primary operating system* and treat the others as limited modules.

Examples:

- A Fortress investor can run a small Harvester sleeve: covered calls on a subset of mature holdings when premiums are attractive.
- A Harvester investor can add a small Trader module: a capped put-selling program on highly liquid names with strict collateral discipline.
- A Trader investor can borrow Fortress ideas: a real spending buffer and a stricter commitment to cash segregation so activity never threatens life cash.

The goal is not to do everything. The goal is to do a few things cleanly, with rules you'll actually follow.

Actionable takeaway: pick your archetype using a two-column decision and write one rule you refuse to break

Make a two-column note right now. Column A: what you want (income level, involvement, flexibility). Column B: what you can reliably provide (time, attention, emotional bandwidth, willingness to be assigned).

Then choose:

- If you value simplicity and stability most, run *The Fortress*.
- If you want steady activity and a balanced income blend, run *The Harvester*.
- If you want maximum income potential and can manage actively with strict rules, run *The Trader*.

Finally, write one non-negotiable rule that matches the archetype:

- Fortress: *"My spending buffer never drops below ______ months."*
- Harvester: *"I only sell calls at strikes I would happily sell the shares."*
- Trader: *"My total put-assignment notional never exceeds ______% of my portfolio."*

If you can keep one rule when you're stressed, you can keep ten when you're calm. The right archetype makes that first rule feel natural instead of heroic.

Chapter 7

Risk Management and Resilience

Building a high-output Income Factory is exciting, but keeping it running during a hurricane is the real challenge. Many investors focus entirely on the 'yield' throttle without checking the 'risk' brakes. In this chapter, we step away from the profit charts to inspect the safety systems. We'll learn why standard stock market metrics fail for option sellers, how to survive the 'gap' risks that happen while you sleep, and why the biggest threat to your portfolio's survival might just be the person looking back at you in the mirror.

7.1 Reading the Safety Gauges: Metrics That Matter

The most dangerous sentence in an income portfolio is: "I'm up on 27 out of my last 30 trades." It sounds like competence. It feels like safety. And it can be the exact pattern you'd expect right before a short—volatility strategy gets hurt—because selling options is a little like collecting nickels in front of an occasionally moving steamroller. The nickels are real, frequent, and satisfying. The steamroller is rare, quiet, and brutally indifferent to your win rate.

Traditional stock investing trains you to watch the usual dashboard: beta, standard deviation, maybe a Sharpe ratio if you're the spreadsheet type. Those are decent gauges for a long-only portfolio where risk tends to scale roughly with market moves. But an income "factory" built on covered calls and cash-secured puts is not a normal car. It's a machine with non-linear parts. Sometimes it purrs while it's actually overheating.

The core issue is simple: standard equity metrics measure *wiggle*. Option selling risk is more about *shape*. A portfolio can look calm most days-small profits, few drawdowns-while quietly building exposure to the kind of move that skips right past your comfort zone. The job here is to replace "feel-good" metrics with gauges that warn you *before* the day you learn a new definition of the word "gap."

Why Beta Can Lie to an Income Portfolio

Beta tells you how a stock (or portfolio) has historically moved relative to the market. If the S&P 500 drops 1% and your portfolio drops 1.2%, you've got a beta around 1.2. Useful-if your risk is mostly linear, like owning shares.

Option selling introduces exposures that do not behave linearly. Two portfolios can have the same beta and completely different "bad day" outcomes.

Here's a concrete way to feel it. Suppose you run a covered call portfolio. Most of the time, you're long stock and short a call, which dampens upside and provides some income cushion. The return path can look smooth: premiums arrive, dividends arrive, and the equity curve rises with reassuring steadiness. Beta might even come out lower than the market. Then a fast selloff hits. Your short call doesn't help much (it was there to cap upside, not protect downside), and your long stock does what stock does when fear arrives: it falls, sometimes hard. If you also sell puts for income-another common factory machine-the downside exposure can grow precisely when volatility spikes and spreads widen.

Beta didn't warn you because beta can't see the *curvature*. You don't need a statistic that says "on average you move with the market." You need a set of gauges that say "if the market moves fast, what happens to me *next*?"

That's the mindset shift: from *average behavior* to *stress behavior*.

Gauge #1: Drawdown Potential (Not Volatility)

Volatility is how much your account value moves around. Drawdown is the pain you actually live through: peak-to-trough decline. If volatility is chop, drawdown is the hole you fall into.

Income strategies can look low-volatility for long stretches because many option-selling trades have a high probability of a small profit. That produces a smooth equity curve-until it doesn't. When a loss finally lands, it can arrive in one large, ugly piece, not in the polite daily increments that standard deviation assumes.

A better question than "How volatile is my portfolio?" is:

> *How deep a hole can I plausibly fall into, and how long would it take to climb out?*

Two income portfolios can both yield 12% a year. One suffers a 15% drawdown every few years and recovers quickly. The other eventually takes a 45% hit that changes your behavior, your confidence, and your future position sizing-meaning the recovery can take far longer than the math alone suggests.

Drawdown potential is the forward-looking version of this idea. You don't just record what happened; you estimate what could happen given your current exposures and the market's current "weather."

A practical way to track it is to keep a simple number on your dashboard:

> *Worst-case 1–3 day loss estimate under a fast market move.*

It doesn't need to be perfect to be useful. What it needs to be is *honest*-and updated when volatility rises, concentration increases, or your positions get closer to the money.

(Visual) Drawdown: The Risk You Can't Ignore

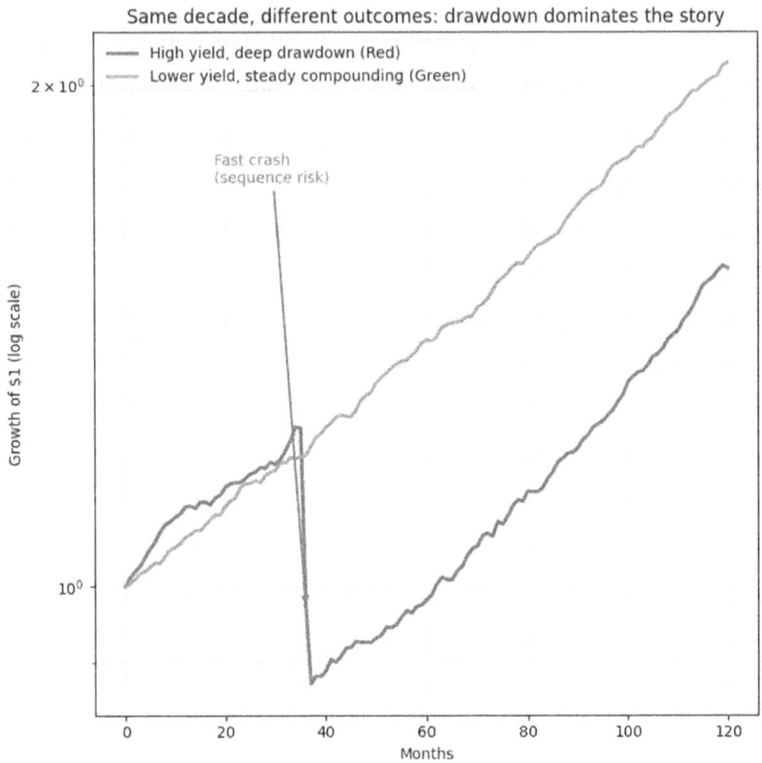

Gauge #2: Delta-How Much Stock Risk You Really Own

Delta is the "stock-like" exposure of your options position. In plain English: delta estimates how much your position's value changes for a $1 move in the underlying.

- Long 100 shares has delta of about +100.
- A short put has *positive* delta (it tends to lose when the stock falls).

- A short call has *negative* delta (it tends to lose when the stock rises), but in covered calls it often just reduces your upside.

The reason delta belongs on your safety panel is that income portfolios often accumulate delta without noticing. A single cash-secured put might look small. Five cash-secured puts across correlated stocks can behave like owning a lot of shares-especially in a down move when deltas tend to rise in magnitude.

Here's the subtle trap: option selling can make you feel diversified when you're actually concentrated. Ten positions can still be one bet if they all load on the same factor-most commonly, "equities don't fall too fast."

A practical delta habit:

> *Track portfolio net delta in share-equivalents and translate it into "How many shares of the index am I effectively long?"*

If your portfolio net delta is +450, that's like being long 450 shares of something. Then you ask the uncomfortable question: *If the market gaps down 3% overnight, do I know what happens to me?*

The goal isn't to run delta at zero. The goal is to run delta *deliberately*. Income investors often think they're managing "trades." Delta forces you to manage *exposure*.

Gauge #3: Gamma-How Fast the Risk Is Changing

Delta tells you where you are. Gamma tells you how quickly "where you are" turns into "where you didn't mean to be."

Gamma measures how much delta changes when the underlying moves. It's the speedometer for nonlinearity. If delta is steering angle, gamma is how quickly the steering locks as the road curves.

Short options are typically *short gamma*. That's the technical way of saying: when the market moves against you, your position often becomes *more* exposed, not less. Losses can accelerate.

This is why a high win rate can be a mirage. Short-gamma strategies often win small and lose big. They're designed to collect time decay (theta), and the trade-off is that your risk can steepen when price moves and when time gets short.

Gamma risk spikes in two common situations:

1. *Near expiration.* Options become "twitchy." Small price moves cause large delta changes.
2. *Near the strike.* At-the-money options carry the most gamma-meaning your exposure can flip quickly.

You don't need to calculate gamma to the fifth decimal. Your dashboard can be simpler:

> *How much of my premium income depends on staying away from a strike in the next few days?*

If your positions are clustered around current prices with little time left, your factory is running hot. It might still be profitable. It's just less forgiving.

Gauge #4: Vega-The ``Volatility Weather" Exposure

Vega measures how much an option's price changes when implied volatility changes. Implied volatility is the market's consensus about how wild the next month could be, expressed through option prices.

Most income strategies are effectively *short volatility*. That means they tend to benefit when implied volatility falls or stays calm and tend to suffer when implied volatility rises-especially during market stress, when volatility can spike sharply.

The emotional danger here is that rising volatility often comes with falling stock prices. So you can get hit twice:

- The underlying moves against you.
- The option premium expands, making your short options more expensive to buy back.

That second effect is the one stock investors aren't trained to expect. In a selloff, your stock position is down-painful, but familiar. In a short-vol portfolio, your "insurance policies sold" also get repriced against you at the exact moment you want them to behave.

A useful macro gauge for many investors is simply the VIX (for U.S. equities), because it's a widely watched snapshot of implied volatility. You don't have to trade VIX products to use it as weather. The point isn't predicting the number; it's noticing regime shifts:

> *When implied volatility is rising fast, your portfolio's risk characteristics are changing fast-even if your positions haven't moved much yet.*

A factory that works beautifully in quiet weather can become over-levered in storm weather without any new trades. Vega is how you see that transformation happening.

Gauge #5: Skew and Tail Sensitivity-Where the Market Is Pricing Fear

Markets don't price "volatility" evenly. They price fear more expensively than optimism. In equity options, downside puts often trade with higher implied volatility than upside calls. That pattern is called *skew.*

You can ignore skew right up until you can't.

If you sell puts for income, skew is the market's way of telling you: "Down moves are the ones that hurt." When skew steepens, it often signals heightened demand for protection. Premiums get fatter, and the temptation to sell them increases-right when the market is shouting that the tail risk is more expensive for a reason.

A practical skew habit doesn't require advanced models. You can do this with what your broker already shows:

> *Compare implied volatility of a 10–20 delta put versus an at-the-money option on the same expiration. If the*

gap is widening, the market is paying up for crash insurance.

This doesn't mean "never sell puts when skew is high." It means treat those premiums as hazard pay, not as free money. Hazard pay should come with smaller size, wider margins of safety, and a plan for what you'll do if the move arrives.

Gauge #6: Liquidity and Spread Cost-Your ``Exit Door Width"

Income portfolios fail in theory because of risk. They fail in practice because of *execution*. When stress hits, bid/ask spreads widen. Liquidity dries up. Your "buy to close" price can be far worse than the mid-price you used in your spreadsheet.

This is why "average P/L" is not enough. You need a gauge for how expensive it will be to unwind your exposures when you *have* to, not when you *want* to.

Watch:

- Average bid/ask spread as a percent of option price.
- Open interest and volume (especially near your strike).
- Concentration in names where liquidity disappears under stress (small caps, distressed companies, meme-like volatility).

A narrow spread is an invisible gift. A wide spread is a tax-paid most heavily during the moments you're already under pressure. Your factory doesn't just need edge; it needs *exits*.

Gauge #7: Margin Cushion and ``House Rule" Risk

Many income strategies are designed to be run with some form of margin or at least with buying power awareness. Even cash-secured strategies can become margin-adjacent if you stack multiple positions, use spreads, or rely on the ability to roll and adjust.

Your broker is not a passive bystander. Brokers can change margin requirements, sometimes quickly, especially during volatility spikes or issuer-specific distress. That means the rules of your game can tighten mid-play.

A margin cushion is the difference between "a bad day" and "a forced day."

Two numbers belong on your dashboard:

> (1) *Buying power used* (%) and (2) *Excess liquidity / available funds*

Then you translate them into a scenario question:

> *If the market moves against me and margin requirements tighten, do I still control my choices?*

Control is the whole point. The income factory is supposed to produce cash flow on your terms. If you're running so tight that an ordinary volatility spike hands decision-making to the broker's liquidation algorithm, you don't have a strategy-you have a hope.

Gauge #8: Concentration by Underlying and by Factor

Diversification isn't the number of tickers. It's the number of different ways you can be wrong.

Many option-income portfolios accidentally concentrate in the same factor exposures:

- Equity beta (stocks fall together in stress).
- Volatility (implied volatility rises together in stress).
- Sector risk (financials in credit scares, tech in rate scares, energy in commodity shocks).

A clean way to monitor concentration is to group positions by:

- Underlying (single-name risk).
- Sector.
- Correlation cluster (the "these always move together" bucket).
- Strategy type (covered call vs short put vs spreads).

Then impose a simple rule that prevents accidental over-commitment. Not because rules are fun, but because panic is expensive.

Putting the Gauges Together: A Small, Practical Dashboard

You don't need twenty dials. You need the handful that change your behavior *before* the loss arrives.

A compact, retail-friendly "instrument panel" might look like this:

Gauge	*What it warns you about (in plain language)*
Worst-case 1–3 day loss estimate	“Could tomorrow erase months of income?”
Net delta (share-equivalents)	“How long equities am I, really?”
Gamma hotspots (near-expiry, near-strike clusters)	“Can a small move rapidly become a big problem?”
Vega / volatility regime (e.g., implied vol rising fast)	“Is the market repricing risk against me?”
Skew (downside fear pricing)	“Is the premium hazard pay—and am I respecting that?”
Bid/ask spread and liquidity	“How wide is my exit door when everyone runs?”
Buying power used / margin cushion	“Do I control my decisions under stress?”
Concentration by factor/sector	“How many different ways can I be wrong?”

Notice what’s missing: win rate. Average monthly income. How many green days you had last quarter. Those are scoreboard stats. They’re not safety gauges.

Safety gauges feel slightly annoying when things are going well because they keep telling you to slow down when you want to speed up. That’s not a bug. That’s the feature.

The Income Factory Mindset: ``Boring" Is a Risk Metric

A strange truth about options income is that the best-run portfolio often feels a little underwhelming. The trades don’t look heroic. The premium collected doesn't make great screenshots. You pass on "juicy" opportunities because the gauges say the machine is already running hot.

If that frustrates you, good. That frustration is your brain noticing the difference between entertainment and resilience.

A factory designed to last doesn't aim to maximize income in the best months. It aims to remain functional in the worst month—because the worst month is the one that gets to vote on whether you keep playing.

Actionable Takeaway: Build a Weekly ``Safety Gauge" Ritual

Once a week (and any time volatility spikes), run a 10-minute dashboard check before you place a single new trade:

1. Write down net delta in share-equivalents and ask: *"If the market drops 3%, am I comfortable with the hit?"*
2. Identify gamma hotspots: near-expiration positions near the money; if clustered, reduce size or extend time.
3. Check volatility regime and skew: if both are rising, treat premiums as hazard pay and cut position size.
4. Look at liquidity: if spreads are widening, assume exits will be expensive and plan accordingly.
5. Confirm margin cushion: decide a minimum "excess liquidity" buffer and treat it like oxygen—non-negotiable.

The goal isn't to predict the storm. It's to avoid being surprised by your own exposures when the weather changes. That's what "reading the gauges" really means: not forecasting the future, but refusing to fly blind.

7.2 Navigating Stormy Weather: Event Risk

A stop-loss feels like a seatbelt. You click it, you relax, you drive faster than you otherwise would. Then the market reminds you of an awkward truth: seatbelts work *during* a crash, not *between* the highway and the cliff.

Event risk is the cliff.

It's the risk of a discrete, scheduled (or semi-scheduled) moment when the price doesn't glide from point A to point B—it *teleports*. Earnings are the classic example: the stock closes at $100, you go to sleep, and it opens at $82 with a new story attached. Dividend dates can do it more quietly, by triggering early exercise and surprise assignment. Macro announcements can do it with a single sentence—sometimes not even a full sentence, just a number that hits the tape and instantly rewrites everyone's expectations.

If the earlier "safety gauges" are your dashboard, event risk is the weather radar. The key skill isn't bravery. It's deciding when to pull the plane into the hangar and let the storm pass.

Why Event Risk Is Different (and Why Option Sellers Feel It More)

Ordinary market risk is the kind you can squint at and manage: the stock drifts down, implied volatility creeps up, you adjust, you roll, you reduce. Event risk doesn't negotiate. It arrives on a clock. It produces gaps—price jumps that happen when the market is closed or liquidity is thin—so your carefully engineered "if it hits $X, I'll exit" plan becomes a nice idea on paper.

Options magnify this problem because option sellers are short insurance. The premium you collect is payment for carrying someone else's fear—and event windows are when fear can become fact. When the move is fast enough, your losses don't increase linearly. They *lurch*. That's why traders who swear they "always manage risk" can still get blindsided: their tools were designed for continuous motion, and the market occasionally moves in discontinuous chunks.

Think of it like this: most days, your portfolio's P/L is a dimmer switch. Event risk days, it's an on/off light.

The Gap Risk: When the Exit Vanishes

Plenty of investors learn the mechanics of a stop-loss before they learn its limitation. A stop-loss becomes a market order once the trigger price trades. That works fine when prices trade *through* the trigger. It fails when the market leaps *over* it.

Here's the nightmare in one frame: you set a stop at $95 because you're "disciplined." The stock closes at $100 on Friday. Bad news hits over the weekend. Monday opens at $80. Your stop-loss didn't protect you at $95 because there was no trading at $95. The order triggers at the open and fills somewhere near $80, depending on liquidity. Discipline didn't disappear—your *assumption* did.

For option sellers, gaps are even worse because:

- Your short options reprice instantly (often with implied volatility expansion on top of the price move).

- Your hedge, adjustment, or roll happens at the worst possible moment: widest spreads, highest emotion, and sometimes reduced margin flexibility.

And now the visual that makes the point in a way words never quite can:

The lesson is not "never use stops." The lesson is: stops are a tool for *continuous markets*. Event risk lives in the discontinuities.

The Three Event Categories That Matter Most

Event risk comes in many costumes, but most of it falls into three buckets that show up again and again in income portfolios.

1) Earnings: The Most Popular Ambush in Finance

Earnings are a scheduled confession. Every quarter, management stands in front of the market and answers—sometimes indirectly—one question: *Were we lying to ourselves about the future?*

Options know this. The market typically "pumps" implied volatility into the expiration that spans the earnings date. That makes premium sellers feel like kids in a candy store: bigger credits, seemingly further-out strikes, better-looking probability-of-profit.

The trap is that the premium is big *because the market expects a jump*. You're being paid to stand on the tracks during the scheduled train time.

Two things make earnings uniquely nasty for short option positions:

- The move can be *directional* and *instant*. A great company can drop 15% on "good" numbers if guidance disappoints or expectations were already too high.
- The volatility "crush" after earnings can help you—unless price overwhelms it. Traders love to talk about selling premium for the crush; the market loves to

remind them that a 20% gap doesn't care about your crush.

A practical earnings rule for an income factory is boring on purpose:

> *If a single earnings event can create a loss larger than the last 6–12 weeks of collected premium in that name, the trade is too big or shouldn't exist.*

That rule doesn't require clairvoyance. It requires respect for the distribution of outcomes.

2) Ex-Dividend Dates: The Assignment You Didn't "Trade For"

Dividends feel gentle. They arrive with the emotional texture of a paycheck. That's exactly why they're dangerous: they can trigger the most annoying kind of risk—operational risk—when you least expect it.

If you are short an in-the-money call on a dividend-paying stock, you are exposed to early exercise by the call holder. Why would they exercise early? Because owning the shares through the ex-dividend date typically entitles you to the dividend. If the dividend is large enough relative to the remaining time value in the option, the rational move for the holder can be to exercise and capture the dividend.

From your side of the trade, this shows up as: *surprise assignment*. You wake up and your shares are called away, or your position changes in a way you didn't schedule, potentially creating:

- Unwanted tax timing (short-term gains when you planned long-term).
- A missing dividend you thought you "owned."
- A new exposure you didn't plan (especially if you were running a covered call as part of a broader structure).

The market didn't do anything evil. You sold a contract that contains rights for the holder. Many equity options are American-style, meaning the holder can exercise before expiration. Your job is to avoid being emotionally surprised by something that is structurally normal.

A simple operational checklist around dividends prevents most of the pain:

> *If I'm short calls on a dividend payer and the call is in-the-money, check the ex-dividend date and compare dividend size to the option's remaining extrinsic value (time value). If extrinsic is small, treat assignment as likely.*

Assignment isn't always bad. Sometimes it's fine. What's costly is *not knowing it's on the calendar.*

3) Macro Announcements: When the Entire Market Reprices at Once

Macro events are the market's group chat. Everyone reads the same headline at the same second, and everyone tries to front-run everyone else's reaction. That's when correlations jump toward one, spreads widen, and liquidity becomes selective.

Common culprits:

- Central bank rate decisions and press conferences.
- CPI, jobs reports, and other high-impact economic prints.
- Major geopolitical escalations that hit over a weekend.

Even if you trade single names, macro events can dominate. A company can have perfect fundamentals and still get dragged by a sudden re-pricing of discount rates or risk appetite.

For income strategies, macro risk often expresses itself through volatility expansion. Your "normal" adjustments can become expensive, and positions that were comfortably out-of-the-money can become uncomfortably close in a hurry.

A good factory habit is to treat macro weeks differently:

> *If there's a known high-impact macro release, avoid stacking new short premium positions that expire right after it—especially near-the-money.*

You don't have to be a macro economist. You just have to know when the stadium lights are about to flicker.

The ``Earnings Volatility Expansion" Trap: Premium That Seduces You Into Gambling

There's a particular kind of temptation that only option sellers really understand. You open the chain, you see inflated premiums, and you feel a little smarter than the market. Your brain says: *"They're overpaying for fear. I'll sell it."*

Sometimes that's true—over long periods, option sellers often harvest a volatility risk premium. The problem is that "often" is not "always," and earnings is one of the rare times the market may be pricing fear with unusual honesty.

The seduction works like this:

1. A stock is quiet most of the quarter.
2. Earnings approaches; implied volatility rises.
3. Selling options suddenly looks like doubling your hourly wage.
4. You size up because the credit feels "worth it."
5. The stock gaps beyond your strike, and the whole quarter of collected premium becomes a coupon you used to buy a loss.

The real mistake isn't selling around earnings; it's confusing *high premium* with *high edge*. High premium can simply be the market correctly pricing a wide distribution of outcomes.

An income portfolio survives by refusing the casino vibe. If the trade makes you feel thrilled, treat that as a warning signal, not confirmation.

Practical Weather Forecasting: How to Spot Storms Before You Sell Premium

Event risk management is mostly calendar work—unsexy, repeatable, effective. The goal is not to predict outcomes; it's to avoid placing your positions directly under the falling piano.

Here are the checks that actually matter when you're about to sell premium:

Know the date, not the rumor

"Earnings soon" is vague. "Earnings after the close on February 19" is actionable. Same for ex-dividend. Same for the Fed.

Treat dates like hard objects. If you can't state the date confidently, you don't have the right to size the trade confidently.

Match expiration to the event on purpose

Many retail losses happen because the expiration date accidentally straddles an event. The trader didn't "sell earnings." They sold a 21-DTE put and forgot that earnings is in 10 days.

There are only a few clean choices:

- Expire *before* the event (you're out of the room when it happens).
- Expire *well after* the event (you give the position time to heal if the first reaction is messy).
- Avoid the underlying entirely during that window (sometimes the best trade is silence).

Notice what's missing: "expire right after the event because premium is juicy." That's the classic way to turn a business into a bet.

Assume correlation spikes during storms

In normal times, diversification across tickers feels real. In stress, it can evaporate. That's why event risk should also be tracked at the portfolio level. Ten positions across ten stocks can still be one macro trade if they all share the same event window (like a Fed decision) and the same exposure (short volatility).

A factory that sells premium across many names must still respect the calendar as a *portfolio* calendar, not a ticker calendar.

What to Do Instead of Trading the Storm

Sitting out feels like failure because it produces no premium and no excitement. That's exactly why it's powerful. The market doesn't pay you for activity; it pays you for being on the right side of risk *most of the time* and still being solvent the rest of the time.

When storm windows approach, alternatives that keep the factory mindset intact include:

- Reducing size (the simplest hedge is humility).
- Moving strikes further away (accepting lower credit in exchange for breathing room).
- Using defined-risk structures when appropriate (spreads) to cap worst-case loss, while remembering that "defined" doesn't mean "small."
- Waiting for the event to pass, then selling premium into the post-event regime when pricing normalizes.

The most underappreciated edge in income investing is the willingness to earn money slowly.

Actionable Takeaway: The ``72-Hour Rule" for Event Windows

Before opening any new short premium trade, run a quick calendar filter:

> *If an earnings release, ex-dividend date, or major macro announcement occurs within the next 72 hours, don't open a new short option position in that underlying (or with that expiration) unless you can explain, in one sentence, why you are being paid enough for the gap risk.*

Write that sentence down. If it sounds like a dare—*"Premium is huge, and it'll probably be fine"*—you're not running a factory. You're chasing weather.

The market will always offer you high premiums right before it tests your risk controls. The professional move is to notice the hurricane on the calendar, close the shutters, and let someone else stand outside holding the umbrella.

7.3 Fire Drills: Stress Testing Your Factory

A good income portfolio can feel like a dependable appliance. You check it, it hums, it spits out cash flow, and you go back to living your life. The problem is that markets don't break like appliances. They break like *systems*: one failure triggers another, and then a third, until the whole thing looks obvious in hindsight and completely unfair in real time.

That's why stress testing matters. Not because you enjoy imagining disaster, but because you enjoy *staying in the game*. A factory that can't survive a bad week is not a factory. It's a fragile trick that works right up until it doesn't.

The goal of a stress test is not to predict the next crash. It's to answer a more practical question:

> *If the market punches me in the mouth, do I still get to make choices?*

When you sell options for income, "choice" usually means three things: you can hold without a margin call, you can adjust without paying panic prices, and you can avoid selling your best assets to cover your worst mistakes. Stress testing is how you rehearse keeping those choices.

Stress Testing Is About *Path*, Not Just Destination

Most investors think in endpoints: "What if the market is down 20%?" That's a headline, not a scenario. What matters is the *path* to down 20%—because the path determines whether you get forced out at down 8% while the market keeps falling, then you watch the recovery from the sidelines with a bruised account and a bruised ego.

Option income strategies are especially path-sensitive because of the machinery inside them:

- Short options can become more sensitive as price moves (gamma risk).
- Volatility can rise as the market falls, inflating option prices against you.

- Buying power can shrink precisely when you want it most.

So a real stress test isn't "down 20% at expiration." It's "down hard, quickly, with volatility spiking, spreads widening, and your broker raising requirements because everyone else is panicking too."

That's the environment where portfolios break. That's the environment you rehearse.

Two Kinds of Fire Drills: Historical Shocks and Homemade Shocks

There are two ways to stress test without pretending you're smarter than the market.

Historical shocks ask: *How would my current portfolio have behaved during something like 2008 or March 2020?* You're not copying the exact moves. You're borrowing the texture: fast drops, gap opens, volatility spikes, correlations snapping tight.

Homemade shocks are blunt but powerful. You pick a set of moves and apply them to today's positions:

- Underlyings down 10%, 15%, 20%.
- Implied volatility up 30%, 60%, 100% (because vol often jumps when fear hits).
- Bid/ask spreads doubled (because "the mid" is a fantasy during stress).
- Margin requirements tightened (because brokers don't enjoy being your unpaid risk partner).

Neither approach needs perfection. The value comes from forcing your portfolio into uncomfortable poses and seeing what snaps first.

A Practical Retail Framework: The "Ten-Point Shock Grid"

Professional margin systems often rely on scenario analysis-evaluate the portfolio at a range of up/down moves and look at the worst loss. You can mimic the spirit of that with a simple grid that fits in a spreadsheet.

Pick a reference underlying for your overall equity exposure (an index ETF works as a proxy even if you trade single names). Then create ten shock points, evenly spaced, across a range that actually scares you. If you're running option income, $\pm 15\%$ is a reasonable starting range for "this could happen faster than my feelings can process."

Now compute (or estimate) the portfolio P/L under each point. If you don't have a model, you can still do a useful approximation:

- Use net delta for the first-order move.
- Add a "volatility tax" if you're short premium (your vega exposure).
- Add slippage assumptions for closing trades.

You're not producing a courtroom-proof valuation. You're building a smoke detector.

Two numbers you want from the grid:

> *Worst-case loss* and *where it occurs* (e.g., down 12% is worse than down 15% because of your strikes).

That second part surprises people. Option portfolios can have "danger zones" where losses peak around certain levels-often near clusters of short strikes. Without a grid, you may never notice that your worst case isn't the dramatic headline move, it's the specific move that pins you right where your risk is concentrated.

The Stress-Test Endpoint That Matters: Margin Call Mechanics

You can survive a paper loss. You can't survive a forced liquidation at the wrong time.

A margin call is not merely "your portfolio is down." It's "your broker's risk model says you may not have enough equity to support your positions." In a fast market, those models can tighten. House requirements can increase. Positions that looked comfortably financed on Tuesday can become "too risky" on Wednesday because the world got louder.

A good stress test therefore ends in a plain-English outcome:

> *After this shock, do I still have excess liquidity / buying power, or do I get a call?*

If the answer is "call," the drill is doing its job. It's telling you that your factory is not designed for bad weather yet.

The most useful stress tests are slightly humiliating. They show you that the portfolio you thought was conservative is actually conservative only in calm seas.

The Hidden Killer: Assignment Cascades

Assignment is not rare in the life of an options seller; it's part of the contract. The problem is not assignment itself. The problem is *assignment timing plus capital usage plus bad luck.* When those stack up, you can get an "assignment cascade"—a chain reaction where one position forces another move, which forces another, until you're selling healthy positions to fund a wound.

Here's a realistic version of the story.

You sell a cash-secured put on a stock you genuinely wouldn't mind owning. The stock drops hard in a broad selloff. Your put goes in-the-money. You're assigned early or at expiration. Suddenly you own 100 shares, and your "cash-secured" cash is now spent.

If that's your only position, fine-you own the shares, you breathe, you move on. But factories usually run multiple machines. If you sold several puts, multiple assignments can land in the same week. Your cash gets tied up. Your buying power shrinks. Your broker gets nervous. Requirements tighten. Then the phone call comes-not from the market, from the margin desk.

Now you're not making the best decision. You're making the *available* decision.

And the available decision is often: sell something liquid and good, because it's the only thing that can be sold quickly. This is how investors end up dumping their high-quality dividend positions at the bottom to cover the wreckage of a short-term options bet that "probably" would have recovered.

The cascade isn't a dramatic villain. It's a bureaucratic chain of obligations.

Here is the flow, cleanly laid out, because seeing it as a mechanism makes it easier to prevent.

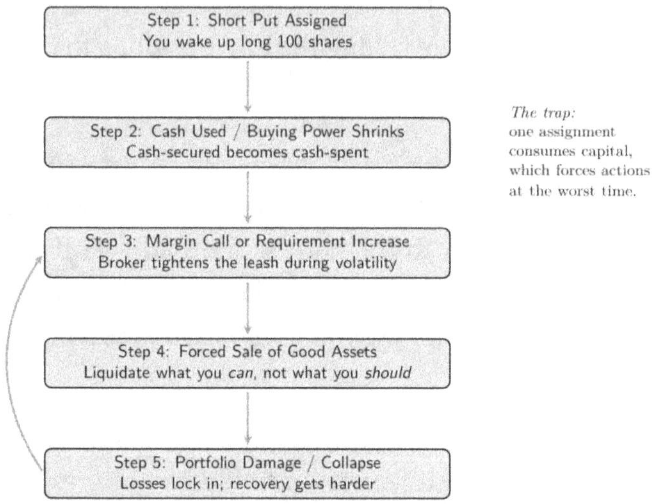

If you've ever wondered why some traders seem "fine" for years and then disappear after one ugly quarter, this is often the mechanism. It's not that they didn't know options risks existed. It's that they never rehearsed the *chain reaction*.

How to Stress Test for Cascades (Without Fancy Software)

To stress test for assignment cascades, you don't need a complex simulator. You need to ask your portfolio a few rude questions.

Rude Question #1: What if my top two puts assign at the same time?

Pick your two largest short put exposures (or the two most correlated). Assume both assign. That means:

- You buy 200 shares total (per contract), at the strike price.
- Your cash requirement is immediate and real.
- Your buying power changes even if prices don't move another cent.

Now ask: *Do I still have enough buffer to keep the rest of the factory running?* If the answer is "barely," the correct response is not "hope they won't both assign." The response is resizing.

Rude Question #2: What if implied volatility doubles before I can adjust?

This is the "the market got scared overnight" scenario. Your short options get more expensive to close. Rolling becomes pricier. Spreads widen. If your plan relies on smooth, cheap adjustments, it's not a plan—it's a fair-weather fantasy.

Rude Question #3: What if my broker increases requirements right when I'm down?

You don't control house rules. You can only control how close you operate to the edge. Stress test by haircutting your buying power: assume you lose a chunk of it instantly. If that creates a margin call, you've learned something priceless while it was still cheap to learn.

Position Sizing as the Real Stress Test

Most stress testing eventually reduces to a single uncomfortable conclusion: you're sized too big for your tolerance, your time horizon, or your broker's patience.

Sizing is emotional because it feels like admitting you're not special. Many traders secretly believe the market will reward their courage. The market rewards survivability. Courage is optional; oxygen is not.

A factory-friendly sizing rule is to cap any single-name "assignment event" so it can't hijack your whole portfolio. In plain terms:

> *No single assignment should consume so much cash that it forces you to liquidate unrelated positions.*

That's not timid. That's industrial engineering.

Make Stress Tests Actionable: Pre-Commit Your Response

A stress test that ends with "wow, that would be bad" is entertainment. A stress test that ends with a specific rule is resilience.

Good pre-commitments are mechanical:

- A minimum cash or buying power buffer you keep *at all times*.

- A maximum percent of portfolio allocated to short puts (or short premium) in any single sector.
- A rule that you do not sell new premium when volatility is spiking *and* your buying power is already strained.
- A rule that any adjustment must be done while you still have choices, not after you're forced.

The point is not to be rigid. It's to avoid improvising under adrenaline, when your brain's only brilliant idea is usually: *"Make the pain stop now."*

Actionable Takeaway: Run a Monthly ``Crash Week" Drill

Once a month, pretend next week is "Crash Week" and answer these on paper before placing new trades:

1. Apply a −15% shock to your core equity exposures and assume implied volatility is up sharply. Estimate your worst-case P/L and, more importantly, your remaining buying power.
2. Assume your two largest short puts assign simultaneously. Write down the cash required and whether it would force sales elsewhere.
3. Assume bid/ask spreads double and you can't exit at the mid. Re-estimate the cost to close your biggest short option positions.
4. Decide *one* concrete change you will make if the drill looks ugly (smaller size, fewer correlated positions, more time, more cash buffer)—and do it immediately, while the market is calm and your pride isn't on trial.

You don't do fire drills because you expect a fire tomorrow. You do them because, when smoke appears, you want muscle memory—not a negotiation with yourself about whether "this time is different."

7.4 The Operator Error: Behavioral Discipline

The market has a sense of timing that feels personal. It waits until you've had a clean month-maybe two-until your watchlist is full of tidy little wins and your account feels like a machine you finally understand. Then it hands you a loss that is not just larger, but *louder*. It takes up more mental space than the last ten winners combined. It follows you into the shower. It sits down at dinner.

That's when the real risk management begins. Not the kind that lives in Greeks and calendars, but the kind that lives in your chest, where you can feel yourself wanting to *do something*. Income investing with options is uniquely good at baiting that impulse, because the controls are always within reach. You can roll. You can add. You can "repair." You can sell another put for "extra premium." You can make the screen change color right now.

Operator error is the moment you confuse activity with control.

The Most Expensive Emotion: Needing to Get Back to Even

Losses don't just hurt financially; they create a psychological debt. "Back to even" becomes a craving-less like a goal, more like relief. And relief is a terrible portfolio manager. Relief will happily trade tomorrow's survival for today's comfort.

This is where income strategies can become dangerous in the most human way. When you're down, a high-yield trade doesn't look like risk. It looks like a ladder out of the hole. Premium becomes emotional currency: each credit is a small hit of *I'm fixing it*. The problem is that the market doesn't care why you wanted that ladder. If you grab the wrong rung-too much size, too little time, too close to the money-you don't climb out. You pry the hole wider.

A useful sentence to keep nearby is this:

> *If I'm changing my process because I'm uncomfortable, I'm not managing risk—I'm negotiating with pain.*

That negotiation usually ends in over-sizing. Over-sizing is how a temporary drawdown becomes a permanent scar.

Loss Aversion: Why a Small Loss Feels Like an Insult

Most people think they fear losing money. That's only half true. What they really fear is *locking in* a loss-the moment it becomes official. Before you close the trade, the loss is still a story you can revise. After you close it, it's history.

This is loss aversion in its most practical form: the pain of a loss is felt more intensely than the pleasure of a gain of the same size. That imbalance is not a character flaw. It's a default setting. The market just happens to be an environment that monetizes default settings.

Options income portfolios amplify loss aversion because they offer so many ways to postpone the "official" moment:

- Rolling a short put down and out.
- Converting a covered call into a collar.
- Selling additional premium to "lower the basis."
- Turning one position into three positions so the original loss disappears into a blur of adjustments.

Sometimes those moves are smart. The danger is when the motive is not expected value but emotional anesthesia.

A simple diagnostic question catches a lot of bad decisions:

> *Would I open this exact trade today if I didn't already own the losing one?*

If the honest answer is no, then "rolling" is not a strategy. It's denial with a transaction fee.

The Rolling Bias: When ``Managing" Becomes Hiding

Rolling is a powerful tool. It's also a hiding place.

The rolling bias shows up when you treat rolls as a moral victory-"I didn't take the loss"-instead of what they really are: a new trade with a new risk profile, placed under pressure, often with worse pricing than you'd choose in calm conditions.

A roll can be healthy when it accomplishes a clear, measurable objective:

- Reduce gamma risk by moving away from expiration.

- Move the strike to a level that meaningfully changes assignment probability.
- Reduce position size (yes, you can roll *and* cut exposure).
- Convert undefined pain into defined pain.

A roll is unhealthy when it has no concrete purpose besides "give it time." Time is not free. Time is a bet that conditions will improve before constraints tighten—margin, liquidity, your own patience.

The real tell is language. If you catch yourself saying, "I'll just keep rolling until it comes back," you're not trading anymore. You're waiting for the market to validate you.

Dopamine Trading: When the Factory Becomes a Slot Machine

The cruel joke of modern brokerage apps is that they make financial risk feel like a video game. There's always a button to press. There's always a chain to scroll. There's always a new expiration to check. The sensation is productivity. The reality is often compulsive behavior wrapped in financial vocabulary.

Dopamine trading is not "trading a lot." It's trading because the act itself is soothing. It's the moment you start using the market the way some people use social media: to regulate mood.

Income strategies are especially vulnerable because they provide frequent feedback:

- Premium credits feel like immediate rewards.
- Many trades show small profits quickly.

- A high win rate creates a sense of competence even when tail risk is growing.

If you've ever found yourself opening a position just to "have something on," you've felt it. The factory is supposed to produce cash flow. Dopamine trading turns it into a noise machine.

There's a disciplined alternative that sounds boring because it is: pre-scheduled trade windows. The factory runs on a shift. Outside that shift, you don't touch the machinery.

Overconfidence: The Quiet Drift From ``Careful'' to ``Cute''

Overconfidence rarely announces itself. It drifts in like humidity.

It starts as small shortcuts:

- "I don't need to check the calendar this time."
- "This is a quality stock; it won't drop that much."
- "I've handled worse."
- "I'll add one more contract; it's basically free money."

Then it becomes "cute" trading-clever structures, too many legs, too many positions, too much complexity for the amount of money at stake. Complexity feels like sophistication. In practice, complexity is often just more places for you to make an emotional mistake.

Overconfidence also tends to spike after a streak of wins. That's when you're most likely to increase size, shorten duration, sell closer strikes, or ignore the signals your own dashboard is flashing.

A factory mindset treats winning streaks as a reason to get *more* conservative, not less. When the machine has been printing smoothly, you're probably underestimating the role of luck, regime, and mean reversion.

The Disposition Effect: Why You Take Profits Early and Let Losers Fester

Many traders do the same strange thing without realizing it:

- They take small profits quickly ("no one ever went broke taking profits").
- They delay taking losses ("it'll come back").

This is the disposition effect, and it's devastating for short-volatility strategies because it flips the logic of the edge. With option selling, your advantage often comes from many small wins. That's fine. But if you systematically cap the winners early while letting losers grow, you turn a high-probability strategy into a low-survival one.

There's a painful irony here: "taking profits early" feels disciplined because it's an action you can brag about. "Cutting a loser" feels like admitting you were wrong. So the ego picks the move that feels like virtue and avoids the move that is actually responsible.

A healthier framing is to treat profits and losses as inventory management:

> *Close winners when the remaining reward is small. Close losers when the remaining risk is large.*

That sentence doesn't care about pride. It cares about asymmetry.

The Recovery Trap: Chasing Yield After a Drawdown

The most common portfolio-killer in income investing is not one bad trade. It's what you do next.

After a drawdown, many investors do some version of "make it back fast":

- Sell premium closer to the money for bigger credits.
- Increase contracts to "earn back" the loss.
- Concentrate in the "best opportunities" (which often means the most volatile names).
- Shorten timeframes because weekly premiums look like a paycheck.

These choices can work in the short term, which is what makes them so dangerous. The market sometimes rewards the behavior just long enough to train you into it. Then the next stress event arrives and you're now larger, closer, and faster-exactly the opposite of what resilience requires.

A drawdown is when the factory should slow down. Most people do the opposite.

Here's the math that makes the point without drama: the deeper the drawdown, the more aggressive you must be to recover—and the more aggressive you must be, the more likely you are to experience another drawdown before you recover. It's a loop.

The way out is not heroics. It's refusing to turn recovery into a deadline.

Discipline That Works: Rules You Can Follow While Upset

The best rules are the ones you can follow when you're tired, annoyed, and slightly embarrassed. If your discipline requires peak serenity, it's not discipline-it's a fair-weather personality trait.

Good behavioral guardrails in an income factory share a few characteristics:

- They are observable (you can tell if you broke them).
- They are mechanical (they don't require a mood check).
- They are protective against the specific ways option sellers blow up: size creep, calendar blindness, and rolling-as-denial.

Below is a concrete set of guardrails you can actually live with. They're intentionally plain. Plain rules survive real markets.

Guardrail: The ``No New Risk When Bleeding" Rule

If you're in the middle of managing a losing position, don't add fresh short premium elsewhere just to feel productive. That's how assignment cascades and margin squeezes become self-inflicted.

Translated into a behavioral rule:

> *If I'm actively defending a position (rolling, taking losses, reducing size), I pause opening new short premium trades until the defense is complete and buying power is stable.*

This stops the common spiral where you sell more premium to finance the pain from selling premium.

Guardrail: The ``One Adjustment'' Rule

Unlimited adjustments are emotional chaos disguised as flexibility. Give yourself a finite script.

For example:

> *I allow one roll for time (to reduce gamma) and one roll for strike (to reduce assignment probability). If the position still threatens the portfolio, I close or reduce it.*

The purpose is not to be rigid; it's to avoid turning one trade into a long-running saga that steals your attention and pushes you toward revenge sizing.

Guardrail: The ``Size Can Only Increase in Calm'' Rule

Most size increases happen during excitement or desperation. Both are bad times to resize.

A surprisingly effective rule is:

> *I only increase my standard position size after a prescheduled review, not during a drawdown and not after a winning streak.*

That one rule protects you from the two classic times humans become reckless: when they feel invincible and when they feel wounded.

Guardrail: The ``Write It Down" Rule

If you have to type a sentence, you slow down. Slowing down is the whole point.

Before any trade that is larger than normal or riskier than normal, write down:

- The specific reason the trade has edge.
- The exact condition under which you will close or reduce it.
- The maximum loss you are willing to accept without negotiating.

If you can't write those in plain language, you're not placing a trade-you're placing a mood.

A Personal Checklist: The Operator's Pre-Trade Safety Card

Pilots don't skip checklists because they're experienced. They use them because they're experienced. They know exactly how the human brain behaves when the engine sputters.

Here's a compact "safety card" you can use before placing a trade, especially after a loss:

Question	*If the answer is "yes," do this*
Am I trying to get back to even quickly?	Cut size in half or skip the trade for 24 hours.
Am I about to sell premium because it feels "juicy"?	Check the calendar; if an event is near, step aside.
Would I open this trade if I had no existing position?	If no, it's probably a roll-as-denial; reconsider.
Is this adjustment increasing complexity or leverage?	Prefer reducing exposure over engineering.
Is my attention being hijacked by one position?	Reduce it until you can think clearly again.

Notice what this checklist doesn't ask: "What do you think the stock will do?" Discipline isn't forecasting. Discipline is preventing your own impulses from becoming portfolio policy.

Actionable Takeaway: Create a ``Red Flag Protocol" for the Next Time You're Tilted

Today-while you're calm-write a one-page protocol for what you do when you're emotionally compromised. Not as self-help. As operational risk management.

Include these three commitments (and keep them brutally specific):

1. *Cooling-off trigger:* Define what puts you on "red flag" status (e.g., any single-day loss over X% of the account, any margin requirement jump, any assignment you didn't expect).

2. *Automatic restrictions:* For the next 48 hours on red flag status, no new short premium positions, no size increases, and no "creative" multi-leg structures.
3. *One productive action:* You may do exactly one thing during red flag status: reduce risk (close, hedge, or cut size). If you can't find a risk-reducing move, you do nothing.

That protocol turns discipline from a personality trait into a machine part. Markets will always tempt you to trade your feelings. The operator's job is to keep the factory running even when the operator is having a very human day.

Chapter 8

Logistics, Taxes, and Continuous Improvement

You've built the machinery, but a factory without a manager is just a pile of expensive scrap. This chapter hands you the clipboard and hard hat, transforming you from a passive investor into an active CFO. We'll tackle the unsexy but profitable art of tax efficiency, establish a stress-free trading rhythm, and learn how to keep score in a way that actually helps you sleep at night.

8.1 Tax Considerations for Income

The fastest way to feel like an investing genius is to look at a yield number before taxes. The fastest way to feel like you're running on a treadmill is to look at your after-tax cash flow and realize a big chunk of your "income" never made it to your checking account in the first place. Taxes don't just reduce returns; they change behavior. They can turn a steady, confidence-building income stream into something that feels oddly fragile—because the government's cut doesn't care whether you *needed* that money for the mortgage or whether the market had a bad month.

The good news is that income investing has a built-in superpower: you get to choose *where* most of this activity happens.

Not which stocks go up. Not whether volatility spikes. But the plumbing—taxable account vs. IRA vs. Roth—often matters more than people expect. Think of it as factory layout. If you place the loud, smoky machines (high-tax income) in the right room (tax-sheltered accounts), the whole operation runs cleaner.

Two tax buckets that quietly run your whole strategy

For most investors, income lands in two basic tax buckets:

- *Tax-friendly income*: typically *qualified dividends* and *long-term capital gains*. These often get lower federal tax rates than wages.
- *Tax-heavy income*: typically *ordinary income* and *short-term capital gains*. These are generally taxed like your paycheck.

That distinction sounds like accounting trivia until you realize it can change your real-world yield by a third (or more) without the market moving a penny.

Qualified dividends are dividends that meet IRS rules (including holding-period requirements and the dividend being paid by an eligible company). When they qualify, they're usually taxed at the more favorable long-term capital gains rates.

Ordinary dividends-common with REITs and some other structures-are generally taxed at ordinary income rates. Not "a little worse." Often *meaningfully* worse, especially for households in mid-to-higher brackets, and potentially worse still if the Net Investment Income Tax applies.

And *options income*? Here's the blunt version that saves confusion: a lot of the cash you collect from selling options in a typical equity-options income approach tends to be treated as *short-term* in nature (or otherwise taxed at ordinary-income-like rates), unless you're using products with special rules (like certain index options) or you're holding positions long enough for long-term treatment. In other words, options can be fantastic cash-flow tools-but, in a taxable account, they often bring a tax bill with teeth.

A quick reality check: gross yield vs. what you keep

A portfolio can "yield" 8% and still feel like 5% once taxes get done chewing. The point of the next chart isn't to give you a universal number (your bracket matters), but to make the *shape* of the problem impossible to ignore: some types of income simply arrive pre-discounted.

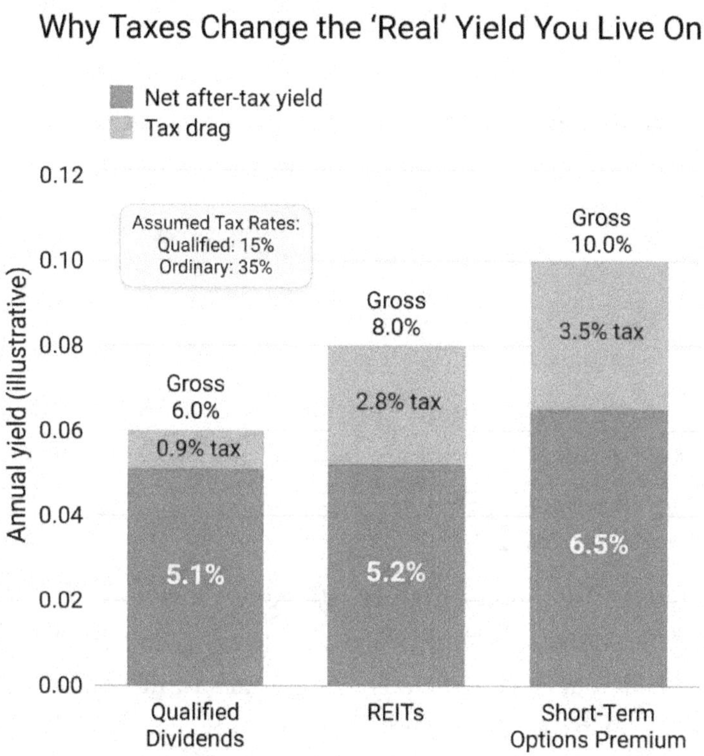

Notice what's really being shown: not "which investment is best," but how the tax system can quietly re-rank them. If two strategies feel similar in risk and effort, the one with less tax drag can win by a mile over a decade-even if its gross yield looks less exciting on paper.

The income types you'll actually run into (and why they're taxed that way)

Qualified dividends: the tax code's version of a coupon

Qualified dividends are the closest thing the U.S. tax code offers to a "thank you for being patient" note. When you own shares long enough (and the payer is eligible), you often get a preferential federal rate.

In practice, this makes mature dividend payers-think many large U.S. companies and broad dividend ETFs-surprisingly efficient *in taxable accounts*. That's counterintuitive for some investors: "Isn't taxable the worst place for income?" Not always. If the income is qualified, taxable can be perfectly reasonable, especially when you also want flexibility: no age rules, no early-withdrawal headaches, and easy access to cash when life happens.

Two practical gotchas keep qualified dividends from being "automatic":

- *Holding period rules*: sell too quickly around the dividend and you may convert a qualified dividend into a non-qualified one.
- *Fund-level complexity*: some funds distribute a mix-qualified, non-qualified, capital gains-so the 1099 matters more than the marketing brochure.

REIT distributions: "income" with a stronger aftertaste

REITs are built to pass through income. The trade-off is that much of what they pay out typically shows up as ordinary income on your tax return. That can make REIT yields look like a feast—until you remember the IRS brought a plate.

There are wrinkles (REIT distributions can include return of capital or other components; some investors may get a deduction on certain REIT dividends depending on current law), but the high-level lesson remains: REIT income often behaves like "high-octane cash flow" that burns hottest in a taxable account.

This doesn't mean REITs are bad. It means they are *honest*. They pay; you pay. If your goal is to maximize spendable income, the question becomes less "Should I own REITs?" and more "Where should I put them so they stop leaking?"

Options premium: a cash machine that may send you a tax invoice monthly

Selling covered calls or cash-secured puts can feel like installing a cash register on your portfolio. Premium hits your account with satisfying regularity. The tax code, however, doesn't grade on satisfaction.

In a taxable account, many common outcomes for option premium translate into short-term gains (or otherwise ordinary-rate treatment), especially when you're harvesting premium frequently. Even when a trade ends as a capital gain, if your holding period is short, the tax rate may look a lot like your wage rate.

Options also create a second tax headache that dividend investors can largely ignore: *they can change the story of your stock position*. A call that gets assigned can accelerate a sale. A put assignment can create a new stock lot with a basis that includes option premium. A roll can turn one simple trade into a string of linked events that are perfectly normal for an options investor and perfectly annoying at tax time.

If you're doing this in an IRA, the day-to-day tax accounting mostly disappears. If you're doing this in a taxable account, you can absolutely do it well-but you're signing up to run payroll.

Asset location: put the ``tax loud" stuff in the soundproof room

Asset location is not about what you buy. It's about *where you hold it*. Two investors can own the same set of holdings and end up with very different after-tax cash flow simply because one used the right containers.

A clean, practical rule of thumb:

- Put *tax-heavy income* (often REITs and frequent options premium) in *tax-advantaged accounts* when possible.
- Put *tax-friendly income* (often qualified-dividend stocks/-funds) in *taxable accounts* when you want flexibility and tax efficiency.

Why this works is almost embarrassingly simple: the tax-advantaged account acts like insulation. You're not "avoiding" taxes forever (traditional accounts defer; Roth accounts may eliminate tax on qualified withdrawals), but you are stopping the strategy from being taxed at the highest rate in the same year you produce the cash.

There's also a psychological dividend: when options and REIT income aren't constantly shrinking after taxes, you stop feeling like you need to take extra risk to "make it worth it." Many blown-up income plans start with a completely reasonable goal and end with somebody selling way-too-aggressive calls because they're trying to outrun tax drag.

A few account-by-account realities (no fantasy settings)

Taxable brokerage: flexible, transparent, and often the best home for qualified dividends

Taxable accounts get a bad reputation because they're honest. Every year, you see what happened. That's not a flaw; it's clarity.

Taxable is often an excellent home for:

- qualified-dividend payers you plan to hold,
- tax-efficient equity funds/ETFs,
- slower-turnover option activity (if you're deliberate about it).

Taxable becomes less comfortable when:

- you generate lots of short-term gains,
- you trade frequently and create wash sales,

- your income pushes you into additional taxes (for higher earners, the extra 3.8% Net Investment Income Tax can apply in certain cases).

Traditional IRA / 401(k): the "factory floor" for high-output, high-tax strategies

If you sell options regularly, a traditional IRA can feel like a cheat code: no annual tax bill on each premium collection, roll, or assignment. The strategy gets to compound without losing a slice every year.

The trade-off is that withdrawals are generally taxable as ordinary income, regardless of whether the original growth came from qualified dividends or long-term gains. That sounds like a loss—until you compare it to paying ordinary rates *every year* on short-term activity in taxable. Deferral is powerful, even when the eventual withdrawal is ordinary.

This is why many investors place their "messiest" income engines here: REITs, option-writing strategies, and other high-turnover approaches that would otherwise spray taxable events all over the calendar.

Roth IRA: the clean room

Roth space is precious because qualified withdrawals can be tax-free. The instinct is to put your highest expected-return assets there, and that can be sensible. But there's another angle: Roth is where you want to avoid "tax regrets."

If you have a strategy that produces a lot of income that would otherwise be taxed heavily (again: frequent options premium, high ordinary distributions), Roth can be an extremely elegant home-if you can tolerate the risk and you respect position sizing. The point isn't to gamble inside a Roth; the point is to keep a good strategy from being sandbagged by taxes for decades.

Options-specific tax potholes that surprise smart people

Options are not "hard" because the math is hard. They're hard because the *edge cases* show up in real life-right when you're busy, the market is moving, and you're trying to keep a calm head.

Assignment isn't a tax disaster; it's a tax event

When a covered call is assigned, you sold your shares. When a cash-secured put is assigned, you bought shares. That sounds obvious, but the surprise is how quickly it can turn a tidy plan into a different tax outcome than you expected.

Example: you're holding a stock with a large unrealized gain in taxable. You sell a call for income. The stock rallies hard, and the call gets assigned. Congratulations-you made money. Also, you may have just accelerated a large taxable gain into *this* year instead of "someday later," and "someday later" might have been a year when your income was lower.

This is not a reason to fear assignment. It's a reason to *choose strikes like an adult*. If you're writing calls on a position with a big embedded gain in taxable, you're not just renting out upside-you're potentially pulling forward a tax bill.

Holding period and "dividend capture" games can backfire

A stock's dividend can look like easy money, and the internet is full of strategies that try to harvest it. The tax code is not impressed. Selling too quickly around the ex-dividend date can cause dividends to be non-qualified. Worse, options overlays around dividends can create outcomes where the economic result is fine but the tax result is strangely disappointing.

The practical response is boring, which is exactly why it works: if you want qualified dividends, hold eligible stocks long enough and avoid cute timing games that turn a tax-friendly cash flow stream into ordinary income by accident.

Wash sales: not a penalty, but a time machine (that moves losses forward)

Wash sales are misunderstood because people treat them like a fine. They're usually not. A wash sale generally *defers* a loss: the disallowed loss is added to the basis of the replacement position, pushing the tax benefit into the future.

The reason wash sales matter to income investors using options is that the "replacement" can involve more than buying the same stock again. Certain option activity can also trigger wash sale treatment if it effectively replaces the position within the IRS window.

What to do with that?

- If you trade frequently in taxable, expect wash sales to happen and plan for recordkeeping.

- If you want clean tax losses, don't immediately hop back into substantially identical exposure just because you feel "naked" without the position.

A deferred loss is not the end of the world. But it can create a year where your taxable income looks higher than your economic reality-exactly the kind of mismatch that makes investors feel like the system is rigged. It isn't rigged; it's paperwork with sharp corners.

Index options and special rules: powerful, but don't assume

Some products (notably certain broad-based index options) can receive different tax treatment than single-stock options. That can be a big deal for active premium sellers who are thoughtful about product selection.

The catch is that this is the part of the map where a casual assumption gets expensive. Product type, where it trades, and how it's classified matter. If you're going down this road, confirm treatment with a tax professional or high-quality tax resources before you scale it. The goal is to build an income factory, not a surprise correspondence course with the IRS.

How to think like a tax-efficient income engineer (without becoming a tax nerd)

A useful mental model: every dollar your portfolio produces has two labels.

- *How reliable is it?* (dividends from a mature company vs. premium from volatile markets)

- *How taxable is it?* (qualified vs. ordinary/short-term)

You want reliability everywhere. But the high-tax label tells you where to place the activity.

A practical "factory layout" that often works well:

- *Taxable account*: mostly qualified-dividend holdings and tax-efficient equity exposure; option activity only if you're deliberate about turnover and the tax consequences of assignment.
- *Traditional IRA/401(k)*: REITs and systematic option-selling (covered calls, cash-secured puts) where you want to reduce annual tax friction.
- *Roth IRA*: selected high-conviction holdings or income engines where tax-free compounding matters most and you can manage risk responsibly.

There are exceptions. If someone is in a very low tax bracket, taxable option income might not be painful. If someone expects higher future tax rates, Roth may be prioritized differently. But as a default, this layout prevents the most common, avoidable mistake: running your noisiest, highest-tax machinery in the least protected room.

The final twist: the ``best" tax plan is the one you'll actually follow

A perfectly optimized plan that's too complicated will fail in the real world. People miss estimated taxes. They forget why a position is where it is. They override the system during a volatile week, then spend the next year cleaning up the consequences.

So aim for a plan you can explain in one breath:

> "Taxable gets my tax-efficient dividend payers. Tax-advantaged accounts get the high-tax stuff like REITs and frequent option premium."

That one sentence can protect years of compounding.

Actionable takeaway

Pick one hour this week and do a "tax drag audit" on your income streams:

- Look at last year's 1099 forms (or your brokerage tax summary) and write down how much of your income was *qualified* vs. *ordinary/short-term*.
- Circle the most heavily taxed category.
- Make one concrete relocation decision: move one high-tax holding or strategy (often REITs or systematic option-writing) into an IRA/Roth if you have space, and move one tax-efficient qualified-dividend holding into taxable if it belongs there for flexibility.

Do that once, and your income factory starts producing the same cash flow with less leakage-no new market risk required.

8.2 Execution Workflows

A good income strategy can die of something embarrassingly small: a sloppy order. Not a thesis failure. Not a recession. Just one panicked click that turns "steady cash flow" into "why did I pay *that* price?" Execution is where your beliefs meet the real world's bid/ask spreads, partial fills, and Monday-morning mood swings. It's also the part of investing that feels least glamorous—so it's where most people improvise. Improvisation is expensive.

The mental shift that matters is treating your portfolio like a business with operating procedures. You're not trying to predict the market every day; you're trying to run a repeatable production line. When the process is consistent, results become diagnosable. When the process is chaotic, you can't tell whether a bad month was the market's fault or your own finger slipping on a market order.

The weekly cadence: decisions when you're calm, execution when you're focused

Income investing with options has a rhythm that rewards planning. The cleanest rhythm is deceptively simple:

- *One day to think*: a weekly scan and plan when markets are closed and your nervous system is quiet.
- *One day to do*: a short execution window when markets are open and liquidity is real.
- *A tiny daily check*: not to tinker, but to manage exceptions.

Why not make decisions in the moment? Because "in the moment" is where your brain confuses urgency with importance. A red candle feels like danger. A green candle feels like opportunity. Both feelings are terrible portfolio managers.

A weekly cadence also fixes a subtle problem: if you look at your account every day without a schedule, you'll end up trading your emotions, not your plan. A schedule turns your attention into a tool instead of a compulsion.

Sunday scan: where you make your best decisions without the adrenaline tax

The best time to decide what you'll do on Monday is when the market can't tempt you. Call it a Sunday scan even if you do it Saturday night. The point is that the quotes are frozen, the news can't whip you into a same-minute trade, and you can actually think in full sentences.

A useful scan has three layers, moving from "big picture" to "specific orders."

Layer 1: Portfolio posture (are you accidentally leaning?)

Start with posture, not positions. Posture is the sum of your exposures-how much of your portfolio is acting like "equity risk," how much is sitting in cash or cash-like holdings, and how much is tied up in option obligations.

Two quick questions catch most problems early:

- *If the market drops 5% this week, what breaks?* (Margin pressure? Too many puts? A concentrated position?)

- *If the market rallies 5%, what do I regret?* (Calls too tight on a core holding? No participation?)

This isn't about forecasting. It's about stress testing your own construction. Many "unexpected" drawdowns are just portfolios discovering what they were all along: more concentrated, more leveraged, or more short-volatility than the owner realized.

Layer 2: Calendar risk (the stuff that makes fills ugly)

Next, look at the calendar. Not because you need to react to every event, but because some weeks have sharp edges:

- earnings weeks for stocks you write options on,
- ex-dividend dates (which can affect assignment risk for calls),
- macro events where spreads widen and prices gap.

You're not trying to be clever here. You're trying to avoid placing yourself in the worst possible microstructure. Selling a covered call into an earnings announcement with a tight strike can be a perfectly valid decision—but it should be deliberate. If it's accidental, it's usually regret waiting to happen.

Layer 3: The "work order" list (exactly what you will place, and why)

Now build a short list of potential trades. Not a brainstorming session—an actual work order.

For each candidate, write down:

- the role it plays (income harvest, entry via put, exit via call, repair/roll),
- what would make you *not* place it (spread too wide, volatility too low, stock moved too far),
- your preferred price (or credit) and your "walk-away" price.

That last bullet is the difference between a professional and a tourist. Tourists click until something fills. Professionals know the price they want and can walk away when the market won't offer it.

Monday execution: a checklist that prevents the dumb mistakes

Execution day is not for debating your philosophy. It's for placing the orders you already designed. You want your Monday session to be short enough that you don't get bored—because boredom is when you start "just checking something," which is how unplanned trades are born.

Here's a practical execution checklist that keeps you out of trouble:

1) Trade in the right hours (liquidity is a real thing, not a vibe)

The first and last minutes of the trading day can be chaotic. Spreads can be wider, pricing can be jumpy, and the "fair" mid-price can be more theoretical than real—especially on single-stock options.

You don't need to avoid the open forever; you need to avoid placing complicated multi-leg orders when the market is still clearing its throat. For many retail investors, mid-morning tends to offer cleaner liquidity. The goal is simple: fewer weird fills.

2) Use limit orders like your returns depend on it (because they do)

A market order is a blank check. It says, "Fill me now, at whatever price the market is offering." That can be fine in extremely liquid products, but with options it's often a donation to the spread.

A limit order says, "Fill me only at this price or better." You're choosing price certainty over speed. That's exactly what an income investor should do, because your edge is usually measured in small, repeatable wins. Slippage—getting filled worse than you expected—turns those small wins into a leaky bucket.

One practical habit: start at the mid-price and "work" the order in small steps. If you're selling a call and the bid/ask is $1.00/$1.20, you might start at $1.10. If you don't get filled, you adjust—slowly. This is not stubbornness; it's refusing to overpay for impatience.

3) Check the bid/ask spread before you fall in love

Before placing any options order, look at the spread. A wide spread is the market telling you, "Price discovery is messy here." You can still trade it, but you should demand better compensation.

As a rule of thumb, if the spread is a large percentage of the premium, you're fighting uphill. That fight shows up as:

- worse fills on entry,
- worse fills on exit,
- and more temptation to use market orders "just to get it done."

You don't want "done." You want "done at a fair price."

4) Confirm contract details like you're defusing a bomb

Options have two common error modes: wrong strike and wrong expiration. Both are easy to do, both are painfully avoidable, and both can instantly convert a careful plan into something you never meant to own.

Before you hit submit, read the order ticket out loud (seriously):

- underlying symbol,
- call vs. put,
- expiration date,
- strike,

- quantity (number of contracts),
- credit/debit and limit price.

This takes ten seconds. It can save you ten hours of cleanup and ten months of regret.

5) Use time-in-force intentionally (your order doesn't need to haunt you)

Time-in-force (TIF) is how long your order stays active. The common choices are:

- *Day*: expires at the end of the trading day if not filled.
- *GTC* (*Good 'Til Canceled*): persists until filled or canceled (often with a broker-imposed maximum).

Income trades often benefit from *Day* orders because you're price-sensitive and conditions change. A stale GTC limit order can fill days later in a different volatility environment—right when you're not paying attention. That's not "automation"; that's setting a trap for your future self.

GTC can be excellent for exits (like profit-taking buy-to-close orders) where you *want* the market to come to you. The trick is to use it consciously, not by default.

Order types that matter most for income strategies (and how they fail)

Most mistakes come from using the right tool in the wrong situation.

Market orders: fast, but you pay for speed

Market orders are great when you truly need out *now* and the product is liquid enough that "now" won't be a robbery. With options, "liquid enough" is not a given.

A market order can fill at a surprising price if the spread is wide or the market moves while your order is routing. If you've ever looked at a fill and thought, "That can't be right," it was right. It was just expensive.

Limit orders: controlled, but you risk not getting filled

Limit orders are the default for disciplined execution. Their failure mode is psychological: you set a limit, it doesn't fill, and you feel rejected. Then you chase.

The professional response to a non-fill is not chasing—it's reevaluating. If the price isn't there, maybe the trade isn't as good as you thought. Non-execution is sometimes the market saving you from yourself.

Stop orders: protection that can become a market order at the worst moment

A stop order triggers when price hits a level, then typically becomes a market order. That means you can get filled in a fast move at a price far worse than your stop level—especially in gaps.

Stops can be useful, but for income strategies they're often less effective than position sizing and thoughtful structure. If you're using stops, understand what you're actually buying: a chance of getting out, not a guaranteed price.

Stop-limit orders: price control, but you may not exit

Stop-limit orders trigger and then become limit orders. You control the price, but you can miss the exit entirely if the market moves past your limit. That's a trade-off, not a flaw. Just don't tell yourself you "have protection" if your protection is optional.

The emotional errors that destroy income strategies (and the antidotes that actually work)

Emotional trading errors are rarely dramatic. They're repetitive. They're small. They feel justified in the moment.

Error: "I need to make back what I missed" (revenge trading in a suit)

Income investors do this when they see a stock run away after their covered call capped the upside. The next move is often writing a higher-strike call too aggressively, or selling extra puts "to catch up."

The antidote is reframing: a covered call that gets assigned is not a failure. It's a trade that did its job. The goal was income plus a defined exit price. If you're consistently furious when you get assigned, your strikes are telling the truth: you didn't actually want to sell.

Error: "This premium is tiny, but it's better than nothing" (income starvation)

When volatility is low, premium shrinks. That's the market saying, "There's less payment available for taking risk right now." Many traders respond by reaching—selling closer strikes, extending duration too far, or concentrating positions.

The antidote is having a minimum acceptable pay rate. If the job doesn't pay enough, the factory slows down. That's not laziness; that's discipline. Cash is a position. Patience is a strategy.

Error: "I can't close it now; I'd be locking in a loss" (loss aversion dressed as principle)

Rolling can be smart. Refusing to take a loss can be catastrophic. The danger is rolling mechanically to avoid admitting a mistake, turning one manageable loss into a long-running obligation.

The antidote is a pre-committed rule: before you open the trade, define what "too wrong" looks like. Maybe it's a delta level. Maybe it's a loss multiple of the premium collected. Maybe it's a technical break on the underlying. The specific rule matters less than having one.

Error: "I'll just manage it intraday" (the illusion of control)

Watching every tick feels like control. It often produces the opposite: overreacting to noise. Options prices can swing wildly intraday even when the underlying hasn't moved much, simply because implied volatility is breathing.

The antidote is scheduled check-ins and alerts for exceptions. If you need to monitor every second, the position is likely too large or too risky for your temperament. Your portfolio should serve your life, not consume it.

Confirmations and reconciliation: the boring habit that catches real money

After a trade executes, broker-dealers send confirmations for a reason: they're a receipt, and receipts matter. Confirmations let you verify:

- the exact price and quantity,
- commissions and fees,
- whether your order filled as expected (or partially filled),
- and whether anything about the execution looks off.

Make it a rule: if you place trades on Monday, you reconcile on Monday. Same day. Not because you're paranoid—because the correction window is larger when the trade is fresh and the details are clear.

This matters even more in a faster settlement world. The U.S. moved to a T+1 standard settlement cycle for most securities, which compresses timelines and makes it more important to catch mistakes quickly. A small error you used to notice "sometime later" can now collide with cash management sooner than you expect.

A practical workflow you can actually keep

A workflow fails when it's too complicated. The goal is to create a routine that survives your busy weeks, your travel weeks, and your "I don't feel like it" weeks.

Here is a stripped-down weekly workflow that tends to stick:

1. *Weekly scan* (*20–40 minutes*): posture, calendar, candidate list, target prices.
2. *Execution window* (*15–30 minutes*): limit orders only; no new ideas allowed.
3. *Same-day reconciliation* (*5 minutes*): read confirmations, verify strikes/expirations/prices, log the trade.

Notice what's missing: constant monitoring. The workflow is designed to reduce decision fatigue and prevent the most common execution mistakes, not to maximize "activity."

Actionable takeaway

Before your next trade, write a one-page *Execution Card* and keep it next to your keyboard:

- *Default order type*: limit order at the mid-price; adjust in small steps; never use market orders for options unless exiting an emergency.
- *Pre-flight check*: symbol, call/put, expiration, strike, contracts, limit price, time-in-force.
- *Walk-away rule*: if the spread is wide or the fill requires chasing, skip the trade and move on.

- *Post-flight check*: read the confirmation the same day and log the fill.

If you follow that card for a month, you'll feel the difference in your results—and in your blood pressure. The market will still do market things. But you'll stop paying unnecessary "execution rent" for the privilege of being impulsive.

8.3 Tracking and Journaling

Brokerage statements are impressive documents in the same way a junk drawer is impressive: there's a lot in there, and none of it is arranged to help you make good decisions. You'll get pages of line items, a "total gain/loss" number that whipsaws with the market, and just enough tax language to make you feel like you should pour coffee directly into your eyes. Meanwhile the one question an income investor actually cares about—*how much cash did my factory produce?*—is usually scattered across dividends, option premiums, assignments, and "miscellaneous adjustments" like confetti.

That's not the brokerage being evil. It's the brokerage doing what it's built to do: report positions and transactions. Your job is different. You're trying to build a machine that produces spendable cash flow without letting market mood swings hijack your behavior. For that, you need a different scoreboard.

The simplest way to change your results is to change what you measure. Not because measurement is magical, but because your brain is a loyal employee: it works on whatever KPI you post on the wall. If you post "portfolio value today," it will panic on down days. If you post "cash generated this month," it will start acting like an operator—calm, consistent, and surprisingly hard to knock off course.

The core idea: separate the factory output from the building's appraised value

An income strategy has two moving parts that shouldn't be mashed into one number:

- *Factory output*: cash you actually generated-dividends received and option premium collected (net of buybacks, commissions, and fees).
- *Building value*: the market price of what you own (and owe) right now-your liquidation value.

The market reprices the building every second. Your factory output accumulates more like a paystub. When you blend them together, you get a performance number that can be accurate yet psychologically useless. You can have a great income month in a bad market and your statement will still look like a disaster. That's how good strategies get abandoned at the worst possible time.

This separation is also how you avoid a common lie people tell themselves: "I'm an income investor." Then they check their account value ten times a day and feel personally insulted by the S&P. Your ledger forces you to act like what you claim to be.

Meet the Factory Ledger: your portfolio's profit-and-loss statement

A Factory Ledger is not fancy software. It's a habit. It can be a spreadsheet, a notebook, or a simple template you copy every month. The essential feature is that it tracks cash flow like a business tracks revenue, while still keeping the balance sheet (portfolio value) visible but not dominant.

Think of it as three pages:

1. *Cash Flow Page*: income produced and costs paid.
2. *Positions Page*: what you own, what you've obligated yourself to (open options), and your sizing.
3. *Quality Control Page*: notes on execution, mistakes, and improvements.

If that sounds like work, it is—about fifteen minutes a week if you keep it simple. The payoff is that you stop being surprised by your own strategy. Surprises are what create emotional trading.

The Cash Flow Page: what came in, what went out, and what counts

Here's where investors get tripped up: not all cash movement is "income." A return of capital distribution is cash, but it's not necessarily income in the economic sense. Proceeds from selling shares are cash, but calling them income is like calling a pawn shop a salary.

A clean Cash Flow Page has only a few categories, but they must be honest.

Income inflows (the stuff you're proud of)

- *Dividends received*: actual cash deposits.
- *Net option premium*: credits collected *minus* debits paid to close, roll, or adjust.
- *Interest*: money market yield, T-bill interest, or sweep interest-small, but real.

Note the phrase *net option premium*. This is where many people accidentally overstate their success. They count every premium collected as income, then forget they spent half of it buying back options to manage positions. Your ledger should count the net, because net is what you can spend without slowly cannibalizing the portfolio.

Costs (the leaks you can actually fix)

- *Commissions and fees*: many brokers are low-cost now, but options still carry regulatory fees and per-contract charges at some firms.
- *Assignment/exercise fees* (if applicable): rare these days, but some firms still charge them.
- *Margin interest* (if you use margin): this can quietly become your largest "expense ratio."

Seeing costs as a line item does something powerful: it turns "annoying friction" into "a controllable expense." Most investors can't control the market. They can control whether they donate $0.08 per share of slippage over and over again.

What does *not* belong on the Cash Flow Page

- *Proceeds from selling stock*: that's converting assets to cash.
- *Deposit/withdrawal transfers*: that's funding decisions, not investment output.
- *Unrealized gains*: those live on the balance sheet, not the income statement.

The discipline here is what makes the ledger credible. If you inflate income by counting asset sales, you'll feel rich right up until the day you run out of assets to sell.

The Positions Page: keeping your exposure visible before it bites you

Income strategies can create "hidden" exposure. Covered calls can quietly reduce upside participation. Cash-secured puts can quietly stack up potential share purchases. Rolling can extend your obligations in time. None of that is bad, but all of it becomes dangerous when you lose track.

Your Positions Page should answer one question instantly: *If I get assigned on everything, what do I own?*

That means listing:

- each underlying position and share count,
- each open option contract (strike, expiration, contracts, and whether it's a call or put),

- the "if assigned" share impact,
- and the cash you've effectively committed (especially for puts).

This page is how you stop a comfortable strategy from turning into accidental leverage. Many blow-ups are simply people being surprised by assignment size. A ledger makes assignment boring, and boring is safe.

The Quality Control Page: your edge is often behavior, not brilliance

Options and dividends don't require genius. They require repeatability. Your Quality Control Page is where you capture the small execution truths that brokerage data won't summarize for you.

Write down things like:

- "Chased a fill; spread was too wide; got a worse credit than planned."
- "Sold calls too close before earnings; assignment risk was higher than I admitted."
- "Rolled too early out of discomfort, not because the trade demanded it."

Those notes sting a little, which is why they work. They create memory. Without a record, your brain will rewrite history into a comforting story where the market was unfair and you were reasonable. With a record, you'll see patterns—especially the ones that cost you money.

A concrete template (simple enough to keep, strict enough to matter)

Below is a no-frills monthly ledger layout. Use weekly entries if you trade actively, monthly if you're lighter-touch. The point is consistency.

Month	*Category*	*Amount*
Jan 2026	Dividends received	$
Jan 2026	Net option premium (credits minus debits)	$
Jan 2026	Interest (cash/T-bills/sweep)	$
Jan 2026	Commissions & fees	($)
Jan 2026	Margin interest (if any)	($)
Jan 2026	*Net factory output (spendable cash flow)*	$

Right below it, add two lines that keep you honest:

Snapshot	*Item*	*Value*
End of month	Portfolio liquidation value (mark-to-market)	$
End of month	Cash buffer (uncommitted cash)	$

The psychological magic here is subtle: you always see the market value, but you stop worshipping it. Your eyes learn to look first at output.

Why this changes your behavior during drawdowns

Imagine two investors in a choppy year.

Investor A checks the brokerage app and sees the portfolio down 7%. Their brain hears: "You're failing." They stop selling puts because they're scared. They stop writing calls because they don't want to "miss the rebound." They freeze, which is the worst time to abandon routine.

Investor B checks a Factory Ledger and sees: dividends arrived on schedule, option premium is down a bit because volatility is lower (or up because volatility is higher), and net output for the quarter is still positive. Their brain hears: "The factory is operating." They keep sizing disciplined, keep using limit orders, and keep producing cash flow. They might still reduce risk if posture demands it, but they won't confuse discomfort with danger.

This is not motivational fluff. It's operating psychology. People don't quit strategies because spreadsheets say they should. They quit because they can't emotionally tolerate ambiguity. A ledger replaces ambiguity with a clearer story: output versus appraisal.

Recordkeeping: the unsexy shield against tax-time chaos

A ledger is for your strategy. Recordkeeping is for your taxes and your future self—who will not remember what you did in April when it's February and you're staring at a confusing Form 1099.

There are two truths worth taking seriously:

- Broker reports are helpful, but they are not a guarantee of correctness for *your* tax situation.
- Brokerage transaction history is not a permanent archive. Firms change platforms, prune history, or present data differently over time.

So keep your own copies. At minimum, store (digitally, backed up):

- annual 1099s and year-end tax summaries,
- monthly statements,
- trade confirmations (especially for complex or high-value trades),
- and your own ledger spreadsheet.

If you trade options actively in taxable, adjustments can arise—wash sales, basis corrections, corporate actions, assignment-related basis changes. Even when brokers report cost basis for "covered securities," you may still need to reconcile and correct items on your return. The ledger won't file your taxes for you, but it gives you a narrative and a paper trail when something doesn't match.

A practical retention rule: keep at least several years of records, and longer for anything that affects cost basis for positions you still hold. If that sounds vague, that's because life is messy: you can't predict which old trade will become relevant when you finally sell a long-held holding.

Trade confirmations: the receipt that prevents phantom mistakes

A confirmation is your receipt. It tells you exactly what happened: price, quantity, time, and fees. People skip confirmations because they assume the system is correct. The system is usually correct. "Usually" is not a risk control.

Make confirmations part of your workflow:

- *Same day*: confirm the fill matches what you intended (strike, expiration, price, number of contracts).
- *Same week*: update your Positions Page so your "if assigned" exposure is current.

This is especially important when you roll positions. Rolling can turn into a chain of transactions where one mis-click creates a different exposure than you think you have. Your ledger is where reality gets reconciled.

Tracking basis without losing your mind (the minimum viable approach)

Cost basis is the "starting value" for tax purposes—the number that helps determine gain or loss when you sell. Options can affect basis in ways that surprise people:

- If you sell a put and get assigned, the premium typically affects the effective purchase price of shares.
- If you sell a covered call and get assigned, the premium typically affects the effective sale proceeds.

You don't need to become a tax software developer. You do need a habit: when assignment happens in a taxable account, record the event in your ledger with enough detail that you can later explain it. Date, underlying, shares, strike, and premium. If your tax forms later look odd, you'll have the story.

For stock positions bought in multiple lots, choose a share identification method you can consistently apply (many investors use specific identification when available). The point is not optimization—it's avoiding accidental outcomes like selling the "wrong" lot and creating a tax bill you didn't intend.

A simple metric that tells you if your factory is improving

Investors love complicated metrics because they look scientific. The metric you actually need is humble:

Net factory output as a percentage of starting capital.

If you started the year with $100,000 and you produced $8,000 of net output (dividends + net option premium + interest, minus costs), you ran an 8% "factory yield" regardless of whether the market priced your holdings up or down in the meantime.

Track it monthly and you'll see patterns you can act on:

- Output rising with stable risk often means your process is improving.
- Output rising because risk exploded means you're borrowing from the future.
- Output falling can be normal in low-volatility environments—your ledger will show whether that decline is due to market conditions or execution mistakes.

Your goal isn't maximal output every month. It's sustainable output with survivable drawdowns.

A story your ledger should be able to tell

At the end of any quarter, you should be able to answer—without guesswork:

- "How much cash did I produce?"
- "What did it cost me to produce it?"
- "What risks did I take to get it?"
- "What would I change next quarter?"

If you can't answer those, you're not operating a strategy—you're collecting experiences.

The long-term smoothing effect (what you want to see over time)

The point of tracking income separately is that, over time, the income line should look like a rising foundation. The market-value line will wobble. Sometimes violently. When you stack them, you can see the hidden stability: even a flat market can produce a meaningful total return if the income layer keeps growing.

Place this picture near the end of your tracking system because it's the "why" your ledger exists: it trains your eyes to respect the slow, relentless accumulation of cash flow.

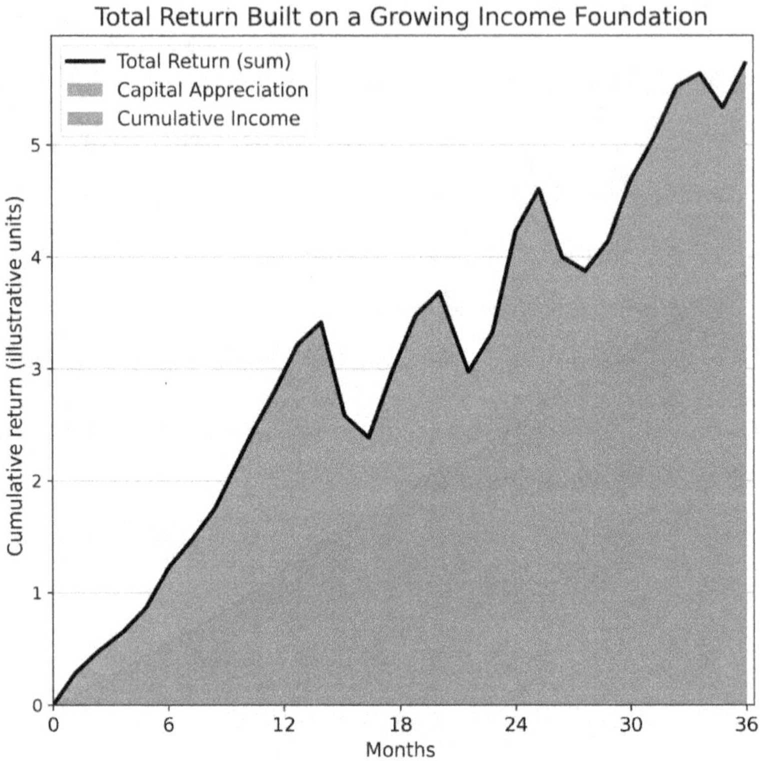

The chart is the visual version of a mental upgrade: even when the top layer shakes, the bottom layer keeps getting thicker. That's what a well-run income strategy is supposed to feel like—less like gambling, more like operations.

Actionable takeaway

Create a Factory Ledger in a spreadsheet tonight and commit to one rule: *every Friday (or every month-end), update net factory output and open obligations.*

To make it frictionless, start with four numbers:

- Dividends received (cash)
- Net option premium (credits minus debits, net of fees)
- Interest earned (cash)
- Total costs (fees, margin interest)

Then add one sentence in your Quality Control Page: the most expensive mistake you made (or avoided) that period.

If you do only that for three months, you'll have something most investors never develop: a strategy you can audit, improve, and stick with—even when markets try to hypnotize you with daily noise.

www.ingramcontent.com/pod-product-compliance
Lightning Source LLC
LaVergne TN
LVHW052024200926
848481LV00005B/1580